HELEN PLUM MEMORIAL LIBRARY
W9-AKL-803

# SILENT SPARKS

## *The Wondrous World of Fireflies*

* * *

SARA LEWIS

PRINCETON UNIVERSITY PRESS
PRINCETON AND OXFORD

Published by Princeton University Press, 41 William Street, Princeton, New Jersey 08540

In the United Kingdom: Princeton University Press, 6 Oxford Street, Woodstock, Oxfordshire OX20 1TR

press.princeton.edu

Jacket art and frontispiece: "Wild Dance of Golden Fairies," courtesy of Yume Cyan / 500px



ISBN 978-0-691-16268-3

British Library Cataloging-in-Publication Data is available

This book has been composed in Perpetua Std text with Trajan Pro display and Scala Sans OT captions

Printed on acid-free paper. ∞

Printed in China

10 9 8 7 6 5 4 3 2 1

For my parents,
who lived and loved together for nearly ninety years,
and fed us wonder
when we were very young.

# Contents

* * *

# PREFACE

* * *

# CONFESSIONS OF A SCIENTIST ENRAPTURED

Fireflies light up our world. Among our most charismatic minifauna, fireflies might just be the best-loved insects on Earth. Time and again, fireflies magically rekindle our sense of wonder. You might cherish some childhood memories of chasing these silent sparks on warm summer evenings. You might be mesmerized right now by their lights twinkling in your own backyard. I readily admit that I'm a firefly junkie, and perhaps you are, too. If you love fireflies, then I wrote this book for you.

I grew up a wild child, enthralled by life's diversity. By the time I'd turned eight I was set on becoming a biologist. Awakened to the natural world by sparkling waterfalls, mysterious hemlock forests, and brilliant night skies, I also vowed to stay familiar with wonder. Little did I know how challenging this would become in my chosen career.

My scientific knowledge burgeoned as I continued my training, first at Radcliffe College, and then at Duke University. For my PhD research I spent years diving on coral reefs to decipher their secrets, occasionally working and sleeping in an undersea habitat sunk 60 feet below the surface. For many decades, my scientific career has been devoted to studying the sex lives of various creatures, including fireflies. As an evolutionary ecologist and professor of biology at Tufts University, I have a truly enviable job: I get paid to be curious, to make new scientific discoveries. I've written hundreds of scientific papers, mentored dozens of students, and won many grant competitions.

And I've tried hard to retain my sense of wonder. But wonder, it turns out, doesn't garner much respect within the realm of academic science. We academics are rewarded for our scholarly productivity—getting research grants and writing technical articles that report our discoveries. Few scientists openly admit to being motivated by wonder. By some unspoken rule, a scientist's feeling of awe for the natural world must be kept under wraps; to acknowledge wonder is tantamount to unreason, and therefore treason. There are some standout exceptions of course, including a few we'll meet in this book.

And scientific reductionism builds its own barriers to wonder. Scientists determine how things work by carefully taking them apart and poking around inside. Consider the miraculous, four-billion-year-old jigsaw puzzle called Life. As scientists, we're trained to look beyond the wonder-inspiring picture displayed on the cover. So we crack open the box, spill out all the pieces. Then we focus down, down, intent on examining each separate piece. Turning each puzzle fragment over and over, we run our fingers along its edges, caressing contours to grasp its shape and deduce its place. Through endless trial and error, experiment and observation, we start to comprehend how these interlocking pieces fit together. With tremendous effort, we might eventually assemble all the pieces to recreate the whole panoramic view. But even then, after focusing so long on the minute details, it's not easy for a scientist to fully recapture that experience of newly minted, breathtaking awe. As the Zen master Shunryū Suzuki put it, "In the beginner's mind there are many possibilities, but in the expert's there are few."

So for me, a scientist working hard to stay susceptible to wonder, this book represents my coming out. I've spent decades delving into fireflies' scientific details, yet these astonishing, radiant creatures still fill me with wonder. Despite my efforts in crafting this book, words can scarcely capture how I feel about fireflies. To me, and probably to you too, fireflies are certainly more than just charming. More than enchanting. Even more than enthralling. The closest I can come: I'm enraptured by fireflies. And delighted, now, that I can finally share the many things I love about them.

It's also a privilege to convey the collective knowledge that scientists have accumulated about these creatures. Fireflies hold so many intriguing tales just waiting to be told! During the past thirty years, researchers from all over the world have divulged some surprising firefly secrets. We've discovered the provenance of their bright lights, learned intimate details of their courtship and sex lives, detected unexpected poisons, treachery, and deceit. Yet the scientific articles describing these discoveries are often written in obscure technical jargon, and are sometimes hidden away behind paywalls. With this book, I've tried to breathe some life into the science and to create an accessible and up-to-date account of these firefly revelations.

I have just one more goal in writing this book. By revealing the wondrous world of fireflies, I'm hoping to persuade people—young and old, in cities and in forests, all around the world—to join me in stepping out into the night. We're surrounded by so many digital distractions it's difficult to find ways to stay connected to the natural world. But we won't need to travel to remote wilderness to experience nature's wonders—these silent sparks are right in our backyards and city parks, just waiting to be discovered.

*Remember—there's no quiz at the end of this book, so read on and enjoy the ride!*

# Silent Sparks

CHAPTER 1

* * *

# Silent Sparks

*And above all, watch with glittering eyes the whole world around you because the greatest secrets are always hidden in the most unlikely places. Those who don't believe in magic will never find it.*

- Roald Dahl -

## A World of Wonder

Fireflies are surely among the most wondrous creatures that share our planet. Living fireworks, these summertime icons fill the night with their spectacular, yet soundless, light shows. For centuries, fireflies' graceful, luminous dances have inspired wonder from poets, artists, and children of all ages. What makes these silent sparks so appealing?

Many of us feel a deep nostalgic connection to fireflies. They evoke childhood memories of summer evenings spent chasing through fields, gathering fireflies in our hands, our nets, and our jars. Peering closely, we marveled at these tiny radiant beings. Sometimes we even squished a few to decorate our bodies and faces with their still-glowing lanterns.

Fireflies create a magic that transcends time and space. Their resplendent displays change ordinary landscapes into places ethereal and otherworldly. Fireflies can transform a mountainside into a living cascade of light, a suburban lawn into a shimmering portal to another universe, a serene mangrove-lined river into a hypnotically pulsating disco.

**FIGURE 1.1** Fireflies spark childhood memories, transform ordinary landscapes, and rekindle our sense of wonder (photo by Tsuneaki Hiramatsu).

All around the world, fireflies elicit a nearly mystical reverence. Surely even the earliest hominids stared awestruck at these silent sparks! Maybe this is what attracts tourists in growing numbers to venture into the night to commune with fireflies. In Malaysia dazzling displays of congregating fireflies draw more than 80,000 tourists each year. In Taiwan almost 90,000 people sign up for firefly-viewing tours during the season. And each June 30,000 tourists visit the Great Smoky Mountains just to admire the Light Show put on by synchronous fireflies. Once I met a woman in the Smokies who'd driven hundreds of miles to see the fireflies there—a yearly pilgrimage she'd been making with her entire family for more than a decade. When I asked what kept them coming back, she pondered a moment and then drawled, "Well, I guess it's just for the awe of it." We all stand in wonder before the silent mystery of fireflies—they move us to joy and thanksgiving.

Fireflies are intricately woven into the fabric of many cultures. But perhaps nowhere on Earth do they shine out through the cultural cloth more brilliantly than in Japan. As described later in this book, the Japanese people have enjoyed a profound love affair with fireflies for more than a thousand years. The still-popular pastime of firefly viewing is deeply rooted in the ancient Shinto belief that sacred spirits, or *kami*, are manifest throughout the natural world. Fireflies became a metaphor for silent, passionate love following the eleventh-century publication of *The Tale of Genji*, a popular novel written by a Japanese noblewoman. Without contradiction, fireflies also came to represent the ghosts of the dead, as powerfully depicted in the 1998 anime classic, *Grave of the Fireflies*. For centuries, fireflies have been celebrated in Japanese art and poetry. These insects feature prominently in many haiku where, like a temporal GPS, fireflies anchor the poem in early summer. Nonetheless, these beloved insects were nearly extinguished from the Japanese countryside during the twentieth century. Yet a remarkable comeback turned fireflies into a symbol of national pride and environmentalism. In Japanese culture, fireflies are like glowing pearls, steadily accreting value with each new layer of symbolic meaning.

## FIREFLY BASICS: WHAT, WHERE, HOW MANY?

Over the past two centuries fireflies have also ignited the spark of scientific inquiry, yielding new insights into their biochemistry, behavior, and evolution. This research really took off in just the past few decades, leading to many exciting discoveries. Beneath their gentle facade, fireflies' lives are surprisingly dramatic—they're full of spurned advances, expensive nuptial gifts, chemical weapons, elaborate subterfuge, and death by exsanguination! This hidden world of fireflies will be revealed in intimate detail within the next chapters of this book.

But first: what exactly *are* fireflies?

These insects go by many different names, including lightningbugs, candle flies, glow-worms, fire bobs, and firebugs. Yet they're neither flies nor bugs—instead, fireflies are beetles. Beetles (also known as Coleoptera) are a famously diverse and successful insect clan. When beetles first evolved 300 million years ago plenty of other insects were already around. But beetles made it big, exploding into a multitude of species. Today 400,000 beetle species live somewhere on Earth, accounting for 25% of all known animals. What gives fireflies their ticket into beetledom? They are all "sheath-winged" insects, their front wings modified into hardened coverings to protect the delicate flight wings.

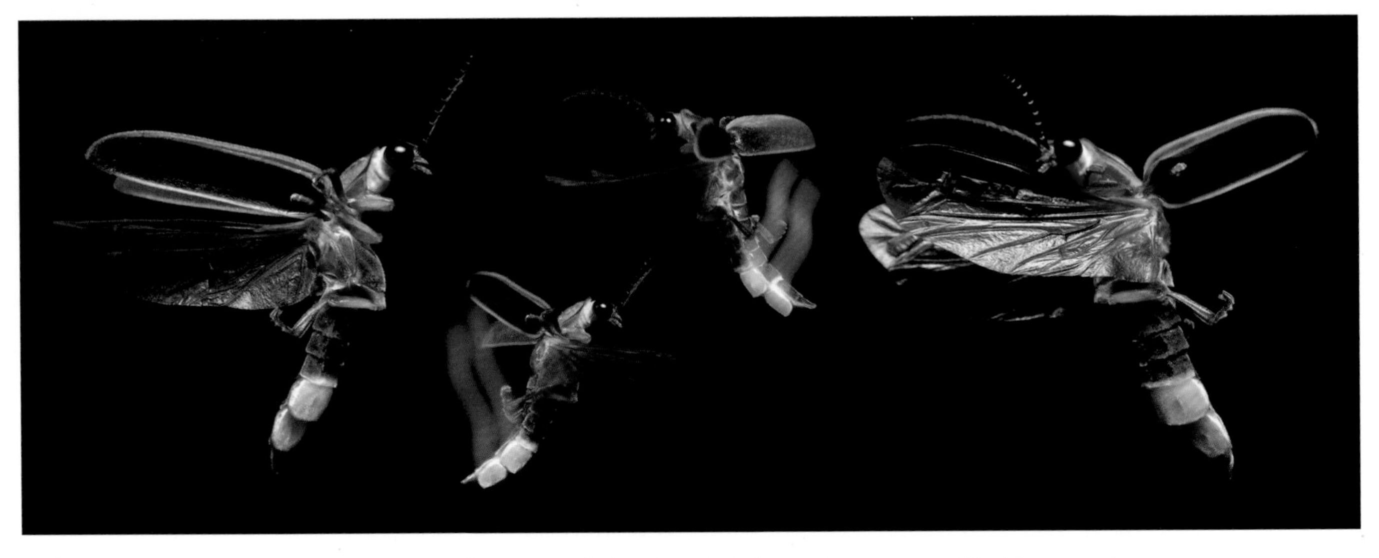

**FIGURE 1.2** Fireflies are really beetles; their front wings are modified into sturdy covers that protect the delicate hindwings, which they use for flying (*Photinus pyralis* photo by Terry Priest).

Within the Coleoptera, all fireflies belong to the family Lampyridae. The beetles in this family are distinguished by several shared features. Bioluminescence (from Greek *bios* for "living" + Latin *lumen* for "light") is certainly one of their signature traits, although many fireflies exhibit this talent only during their juvenile stages. Lampyrid beetles are also distinguished by their relatively soft bodies. If you've ever held a firefly, you might have noticed they're a bit squishy, in contrast with the rigid, shell-like bodies typical of many other beetles. Finally, every firefly prominently carries a flattened shield that covers the back of its head.

All living fireflies not only look alike, but they also trace their genetic roots back to a single common ancestor. This protofirefly probably lived about 150 million years ago, back in the dinosaur-dominated Jurassic period. At that time, insects were spreading out and diversifying to fill new ecological niches (including a cockroach that specialized in eating dinosaur dung!). While we don't know what ancient fireflies might have eaten, we do know that as far back as 26 million years ago, fireflies already resembled the ones we see today. We know this because some fireflies got tangled up in sticky tree resin that later hardened into amber, trapping the creatures inside and preserving them in exquisite detail. One piece, dating from 19 million years ago, contains two fireflies caught in the act of mating, now forever sealed together in love (Figure 1.3).

**FIGURE 1.3** Long ago, these two fireflies were caught in flagrante delicto when they became trapped in tree resin (photo used with permission of Marc Branham).

People are often surprised to learn there are many different kinds of firefly, not just one. In fact, there are nearly 2,000 firefly species sprinkled across the globe. Collectively, fireflies stretch from Tierra del Fuego at 55° south latitude to Sweden at 55° north latitude, gracing every continent save Antarctica. As is true for most living things, firefly diversity rises in the tropics, peaking in tropical Asia and South America: Brazil alone hosts 350 different firefly species. There are more than 120 recognized firefly species in North America; here the greatest diversity occurs in the southeastern United States, particularly Georgia and Florida. These states are each home to about fifty different firefly species, while in all of Alaska there's only one. For many years, the scientific study of fireflies was largely focused on cataloging new species; that is, finding, naming, and formally describing their anatomical differences. Even today, new firefly species are still being discovered.

## Looking for Love with Flashes, Glows, and Perfumes

As fireflies blossomed over evolutionary time, they hit upon remarkably diverse ways to find and attract mates. Present-day fireflies can be conveniently grouped according to their courtship styles: some species use quick, bright flashes of light to locate mates, while others use slow glows, and still others use invisible windborne perfumes.

*Lightningbugs* earn their name from their talent for flashing—both sexes speak their love in the language of light. Justifiably famous for their brilliant nocturnal displays, these are the fireflies most familiar across North America. Their precise on-off flash control allows lightningbugs to carry on elaborate conversations with

**FIGURE 1.4** A typical North American lightningbug; males of the American Big Dipper firefly court their females with quick, bright, *J*-shaped flashes (*Photinus pyralis* photo by Alex Wild).

prospective mates. Typically, flying males broadcast distinctive flash patterns, while sedentary females flash back in response. This courtship style has evolved in several different firefly lineages. Lightningbug fireflies are ubiquitous east of the Rocky Mountains, yet for reasons unknown they're found only in scattered pockets across the western states and provinces.

*Photinus pyralis*, commonly known as the Big Dipper firefly, could easily be the poster child for all lightningbug fireflies. Its common name reflects this beetle's large size (up to 15 mm long) as well as its flash gesture: while flashing, males dip down then sharply rise up, skywriting the letter J with their lights. Big Dipper fireflies are found all across the eastern United States from Iowa to Texas, and from Kansas to New Jersey. Active just at dusk, they fly close to the ground—even young children can easily capture them. And these lightningbug fireflies are not too picky about habitats; they're often seen flying over suburban lawns, golf courses, roadsides, parks, and college campuses.

**FIGURE 1.5** A European glow-worm firefly; clinging to her perch, the female dangles her lantern to attract flying males (*Lampyris noctiluca* photo by Kip Loades).

Northern Europe has mainly *glow-worm fireflies*, whose plump and wingless females give off long-lasting glows. Flightless and earthbound, every night these females clamber up onto low perches and glow for hours to attract flying, but typically unlit, males. Some glow-worm females also add chemical scents to their love potion. Released into the air, such perfumes flow unobstructed around trees and other vegetation, attracting males from afar.

Worldwide, nearly one-fourth of all firefly species are glow-worms. Best-known among them is the common European glow-worm, *Lampyris noctiluca*. Widespread, this particular glow-worm firefly occurs throughout Europe from Portugal to Scandinavia, and across much of Russia and China. The glow-worm style of courtship is also popular in many Asian fireflies. Oddly, glow-worms are rare among North American fireflies, though a few are found even west of the Rocky Mountains.

**FIGURE 1.6** A North American day-flying dark firefly; instead of lights, these adults use scents to attract mates (*Lucidota atra* photo by Peter Cristofono).

Equally widespread are the *dark fireflies*, so named because these adults fly during the daytime and they don't light up. Instead these males locate mates by sniffing out wind-borne perfumes given off by their females. Evidence suggests that the very first fireflies used similar courtship styles. Such day-active dark fireflies are common across North America, even in the West.

### Firefly Semantics

Although fireflies account for *most* of the world's glow-in-the-dark beetles, this bioluminescent ability is shared with beetles from other families, including the phengodid beetles (Phengodidae, or railroad worms) and a few click beetles (Elateridae; called *cucubanos* in Puerto Rico, *peenie-wallies* in Jamaica).

So what exactly do we mean by the name "firefly"? This term refers to any member of the beetle family Lampyridae, whether or not their adults light up. Fireflies can be sorted into three groups according to the different courtship styles they use to find their mates:

- *lightningbug fireflies*: adults use quick on-and-off flashes for courtship
- *glow-worm fireflies*: flightless females produce long-lasting glows to attract males; typically, the males don't light up
- *dark fireflies*: these adults don't light up; they court during the day, relying on chemical cues to find mates

## On the Move

By accident or with intention, people have sometimes transplanted fireflies to places where they don't belong. Lesser glow-worms, *Phosphaenus hemipterus*—native to Europe—were discovered in Halifax, Nova Scotia, in 1947, probably traveling as stowaways in the soil of imported tree seedlings. These flightless glow-worms managed to survive and even spread out around Halifax, where several populations were still thriving in 2009. But other transplants didn't take. Around 1950, some flashing fireflies (*Photuris*) were deliberately introduced from the eastern United States, in hopes of adding some sparkle to city parks in Portland, Oregon, and Seattle, Washington. They flickered for a few weeks, then disappeared. Another transplant attempt during the 1950s introduced Japanese fireflies to control snails in Hawai'i; these transplants likewise failed to survive. No one knows why some firefly transplants failed while others thrived. Maybe the destination address had the wrong temperature, moisture, or soil conditions, maybe some favorite prey was lacking, or maybe some new predator was lurking.

We now recognize it's generally a bad idea to intentionally relocate creatures—even beautiful, apparently harmless ones like fireflies. Many gorgeous plants—like purple loosestrife, water hyacinth, and Japanese knotweed—were originally imported to the United States as ornamentals. But these exotics quickly turned invasive, weedily crowding out native species and wreaking ecological havoc. Every living species is embedded in an elaborate, yet often obscure, network of biological interactions. When we disrupt this network by plucking creatures from one place and plunking them down somewhere else, there's really no telling what will happen.

## What's Up Next

This book offers a guided tour of fireflies' luminous lives. We'll hear behind-the-scenic stories about their courtship rituals, their potent poisons, some seductive mimics, and their present plight. Of course, these remarkable tales would remain untold if a few inquisitive scientists hadn't invested their days and nights to deciphering these firefly mysteries. Beginning with my own seduction into firefly biology in the 1980s, I've been privileged to know and work with many leading researchers. As the fireflies' stories unfold, we'll meet some of the people whose lives are so intimately entwined with these creatures. They're not merely scientists. They include Lynn Faust: horsewoman, mom, and self-taught naturalist whose firefly-spangled childhood inspired her to become the leading expert on the Light Show of the Great Smoky Mountains. Another is Raphaël De Cock, a traveling troubadour who leads a double life, because he is also an expert on glow-worms. We'll head out into the night with James Lloyd, the solitary field biologist who has devoted a lifetime of summer nights to observing how fireflies behave in the wild. And we'll hear about the late John Buck, a dedicated sailor and physiologist whose careful laboratory studies uncovered the mechanisms that fireflies use to control their flashing. Together with others from around the world, these scientists' collective efforts have divulged some of fireflies' most deeply held secrets.

Before we enter this mysterious world, here's a quick peek at what's coming up next:

In chapter 2, "Lifestyles of the Stars," we'll see that all fireflies rise up from humble beginnings. Fireflies enjoy a remarkable childhood. They spend most of their lives—up to two years—as grub-like juveniles that live underground. Baby fireflies turn out to be fearsome predators, indulging themselves in gluttony and growth. We'll follow a firefly through its different life stages, starting from a faintly glowing egg through the curious magic of metamorphosis. Once fireflies reach adulthood, they become single-mindedly focused on sex. Traveling to Tennessee, we'll walk into a forest to find waves of light created by the synchronous pulsations of a thousand fireflies washing over us.

The spectacular displays we so admire are actually the silent love songs of male fireflies. In chapter 3, "Splendors in the Grass," we'll visit a New England meadow at dusk to experience some lightningbug courtship. For nearly three decades my students and I have studied fireflies in the wild to gain insight into sexual selection, a subtle yet powerful evolutionary process. We'll follow some males as they lift off on their nightly search missions. As these flying, flashing males earnestly broadcast their availability, coy females only flash back if they spot an especially attractive male. What exactly does a firefly female consider sexy? We'll find out. And we'll see how steeply the odds are stacked against firefly males. Only a select few will end up embracing an amorous female, while many others will embrace only death.

But what happens after the lights go out? Firefly sex is enigmatic, and not just because it happens under cover of darkness. As we'll learn in chapter 4, "With

This Bling, I Thee Wed," the drama now continues deep within the hidden recesses of females' reproductive system. My own microscopic explorations of fireflies' interior landscapes led to a thrilling discovery that transformed our view of firefly sex lives. In this chapter we'll hear about nuptial gifts, and see what these amorous bundles mean both to the gift givers and to the recipients.

Because glow-worm females are flightless, their lifestyle contrasts sharply with that of other fireflies. In chapter 5, "Dreams of Flying," we'll meet a rare American glow-worm, the blue ghost firefly. We'll tag along on a field expedition to the southern Appalachian Mountains to study the courtship habits of these mysterious blue ghosts. Flying ankle-high above the forest floor, males give off eerie, long-lasting glows as they search for females. Meanwhile, the blue ghost females, tiny and wingless, crawl slowly through the leaf litter. Trying to glimpse the world as they see it, we'll enter the Umwelt of these flightless females, shrinking down to assume their vantage point.

How do these creatures *make* light? Firefly glows appear magical, but in chapter 6, "The Making of a Flasher," we'll learn how bioluminescence arises from carefully orchestrated chemical reactions. Inside the firefly's lantern we'll discover the main star, luciferase, an enzyme that's also been put to work protecting human health. Some fireflies can quickly switch their lights on and off, a talent that allows them to blink out precise, Morse-code-like signals. How do these insects achieve such high-tech flash control? We'll also accompany some early firefly biologists on a quest to Southeast Asia to discover how some fireflies manage to flash together all night long in marvelous synchrony.

But fireflies aren't all sweetness and light. Chapter 7 ("Poisonous Attractions") divulges the dark side of fireflies. Some fireflies manufacture potent poisons, and these nasty-tasting chemicals keep birds, lizards, mice, and other insectivores at bay. These chemical weapons also hold the key to understanding why fireflies' bright lights evolved in the first place. Yet fireflies are considered quite a delicacy by certain creatures, including some flashy and alluring femmes fatales.

The world of fireflies holds many tantalizing tales, and we still have lots to learn. But meanwhile, firefly populations around the world are blinking out. In chapter 8 ("Lights Out for Fireflies?"), we'll explore the complicated, often destructive, relationship between humans and fireflies. We'll take a look at some likely culprits behind recent firefly declines, including habitat destruction and light pollution. We'll also learn how people have overexploited fireflies, sometimes to extract their light-producing chemicals, other times to enjoy their numinous beauty. Fortunately, there's still hope that future generations will be able to enjoy these living sparks. This chapter wraps up with some practical ways to make your yard more firefly friendly.

At the end of the book, we'll gather our gear and step out into the night to get personally acquainted with some local fireflies. A field guide will help you identify some common North American fireflies and learn how to decode their courtship conversations. The final section offers pointers on useful field gear and charts some adventures that will let you peer even more deeply into the many wondrous worlds of fireflies.

In this book I've decided to forgo the usual scientific displays, like graphs and tables. But if you're so inclined, you'll find an annotated bibliography for each chapter that points to the relevant information. When scientific articles are available online without charge, I've provided links to these resources. (In the e-book version, you can click directly through; if you're reading the print version, you'll find all the chapter notes, including these links, on the *Silent Sparks* blog or at http://press.princeton.edu/titles/10667.html.) I've also provided a glossary to help explain some specialized terms. And should you want even more, you'll find a carefully selected list of web and print resources to guide further explorations.

So let's begin our journey into the hidden world of fireflies. We'll be traveling behind the scenes to explore the nightly drama that unfolds in your own backyard, in your neighborhood park, and in the fields and forests nearby. Let's open the door and step lightly through . . .

CHAPTER 2

* * *

# Lifestyles of the Stars

*The only way to make sense out of change is*
*to plunge into it,*
*move with it,*
*and join the dance.*
- Alan Watts -

## Deep in the Heart of the Smokies

A few years back, I was treated to a truly awe-inspiring firefly experience. At the Great Smoky Mountains National Park I met up with Tennessee native Lynn Faust, a remarkable naturalist who openly admits her firefly obsession. That June we joined thousands of visitors who arrive here each year during two carefully timed weeks to watch one of nature's most extraordinary displays. Now wildly popular, until the mid-1990s this spectacle was familiar to only a few families whose rustic summer cabins made up the tiny town of Elkmont.

Faust spent her childhood summers here in what she calls "the magic of Elkmont," roaming misty forests and splashing through trout-filled mountain streams. Faust nearly sparkled as she told me about the evening ritual they followed on certain June nights. Clad in their pajamas, all the neighborhood kids would gather together on the screened porch after dinner. There they eagerly awaited what Faust's future mother-in-law had christened "The Light Show." They watched as darkness descended on the surrounding forest. First they saw just ten, then a hundred, and finally thousands of fireflies, all flashing simultaneously in a silent symphony of light.

These childhood experiences got Faust hooked on Elkmont. Later, she and her husband, Edgar, returned here every year with their young family. When the Great Smoky Mountains National Park was established in 1940, the federal government allowed Elkmont families to remain under long-term leases set to expire on December 31, 1992. That date loomed ominously over half a century, threatening an end to the magic for the Fausts and all the other Elkmont residents. Faust tearfully recalled the night when, at the stroke of midnight on New Year's Eve, armed park rangers arrived to politely but firmly escort her family and friends out of their Elkmont cabins. A few months later, Faust—by now an acutely keen observer of fireflies—contacted Dr. Jon Copeland, a biologist from Georgia Southern University. Copeland had just returned from a far-flung expedition to Southeast Asia to see the famous synchronously flashing *Pteroptyx* fireflies there. Initially he was skeptical about Faust's description of her Tennessee fireflies, because synchronous fireflies weren't supposed to live in the United States. But during the summer of 1993, Faust convinced Jon and his colleague Andy Moiseff to visit Elkmont and see for themselves. Copeland would later laconically describe that wonderful moment of discovery that dawns every now and then on a scientist: "It was foggy, cold and rainy. It got dark and nothing happened. I was sitting in my car, feeling a little sleepy. When I opened my eyes, fireflies were flashing synchronously!"

After their visit, Faust began to spend her summers working as a field assistant with Copeland and Moiseff. Both insect neurophysiologists, the two men were dead set on figuring out how, and why, these particular fireflies maintain their precision synchrony. Now Faust really started paying attention to exactly what the Elkmont fireflies were doing. She kept detailed field notes all summer long. Each winter she'd type these up, assembling them with the photographs she'd taken into neat, spiral-bound volumes, one for each year. Filling bookshelves, Faust's field notes now span more than twenty years. She's also managed to keep close watch on more than a dozen other firefly species that live in eastern Tennessee.

Over the years, more and more visitors have flocked to the Great Smokies to view the natural spectacle created by the courtship rituals of *Photinus carolinus*, the particular firefly species that puts on the main act. Initially, visitors could drive to Elkmont, but the fireflies' increased popularity led to traffic nightmares, while car headlights disrupted the fireflies. In 2006 the National Park Service began providing shuttle bus service between the best firefly-viewing spot and the Sugarlands Visitor Center in Gatlinburg. Now nearly 30,000 visitors arrive during the two-week period in June to admire what is officially dubbed "The Light Show." The requisite shuttle bus reservations, available only online, sell out within minutes every year. In 2008 I met up with Faust to witness this spectacle firsthand, and to study these fireflies' mating behavior.

As Faust and I stepped off the shuttle, the leafy aroma of forest and ferns and the roar of water rushing down the mountain greeted us. Our fellow visitors included church groups, young lovers holding hands, and multigenerational families with kids racing ahead and elders trailing behind. Maybe they'd come to seek communion with their own personal God, or with one Being, or to otherwise connect with something larger than themselves. Whatever they were looking for, they must have found it. Many visitors I spoke with had made this pilgrimage many times before, returning to see the Smokies Light Show year after year.

We walked slowly up the path that winds along the Little River, passing shells of the old Elkmont cabins abandoned by decree and now ignored by the Park Service. We headed down a side trail to peer inside the screened porch of the Fausts' former cabin, where the old dining table was still awaiting its family's return. Elkmont looked eerie in the failing light, its ghostly cabins, pathways, and gardens dilapidated and despondent, all slowly decomposing into ruin.

The crowd began to disperse at the top of the rise, and small groups of visitors settled into forest clearings here and there. They unfolded the lawn chairs they'd carried and sat down to wait patiently as dusk fell. Unlike most crowds, these spectators seemed almost reverent, like visitors to a cathedral. Everyone sat quietly, speaking together only in hushed voices. It wasn't until the forest was completely dark that we saw the first flash. A few moments later a dozen male fireflies took flight around us, broadcasting their typical mating call: six rapid bursts of light followed by six seconds of total darkness. Suddenly, the forest came alive with flying sparks, and thousands of male fireflies were flashing together in lock-step synchrony! Together they flared out their six precisely timed flashes, and then they all ceased at once. Darkness rushed in like a shade drawn over my eyes.

All the scientific descriptions that I'd read had left me totally unprepared for the transcendental thrill of this rhythmic, pulsating display. Mesmerized, I sank down and yielded to this immense, hypnotic biological rhythm. As Steve Strogatz put it in his wonderful book *Sync*: "At the heart of the universe is a steady, insistent beat." Synchrony in any sensory channel is strangely compelling to our human

**FIGURE 2.1** Visitors to Elkmont are greeted by a traveling symphony of lights (*Photinus carolinus* photo by Radim Schreiber).

minds. Alone in the silence save for a synchronous symphony played by a thousand fireflies, I felt like I'd fallen out of time.

That night in Elkmont I witnessed a prodigious effort that, as we'll see shortly, was all about procreation. The tiny, pulsating stars responsible for this brilliant display were desperately trying to propel their genes into the next firefly generation. As for the rest of us, we were just fortunate to be spectators at their exhibition.

When at last I recovered, Faust and I spent the next several hours looking for female fireflies and planning out the observations we'd be gathering over the upcoming nights. It was long past midnight when we finally stumbled down the trail. As we walked out of that magical forest, I marveled at this natural wonder that had lain hidden for so long deep in the heart of the Great Smokies.

## Humble Beginnings

Like all fireflies, the architects of the Elkmont Light Show rise up from humble beginnings to spring forth as future stars. During their life cycle, fireflies undergo an impressive self-transformation known as complete metamorphosis. Originating among insects about 290 million years ago, this complicated lifestyle has proven wildly successful over evolutionary time. Today, every beetle, butterfly, bee, fly, and ant goes through complete metamorphosis; taken together, these creatures account for roughly half of all the animal species on Earth.

Insects are virtuosos of change. Human development doesn't even come close. Our babies, like other mammals, are basically miniature adults—we grow ever larger, but we've still got pretty much the same parts list. In contrast, insects' transformational power is astounding: they completely reinvent their bodies as they grow. Until the seventeenth century, in fact, caterpillars and butterflies were thought to be entirely different creatures.

Metamorphosis not only provides the ability to shape-shift, it also lets insects change how they make their living. Juveniles and adults can now live in different habitats, so they can take advantage of different resources. And they can also specialize in performing distinct tasks. Typically, juveniles dedicate themselves to eating and growing (surviving too, of course). And adults devote themselves to the pursuit of sex, and sometimes to dispersing into new habitats.

A firefly's life is full of contradictions. During its lifetime it will drastically change both its profession and its personality—an insect equivalent of Mr. Hyde becoming Dr. Jekyll. We might admire the adult firefly for its ethereal grace, but it wasn't always such a gentle creature. Fireflies spend their childhood in a radically different juvenile stage called a larva (plural larvae). Firefly larvae are voracious carnivores, capable of subduing and devouring prey several times their size. Unfortunately (or maybe fortunately), you rarely see them because these juveniles live cryptic lives. As larvae, most US firefly species live underground and feast on earthworms, snails, and other soft-bodied insects. The larvae of many Asian fireflies live underwater and feed on aquatic snails. Surprisingly, fireflies spend the vast majority of their lives in this juvenile stage. In northern latitudes, the larval stage probably lasts between one and three years, while farther south it might only last several months. When the time is right, the firefly larva seeks a safe place to become an immobile pupa (plural pupae). This pupal stage lasts about two weeks, during which the firefly is busy rearranging its body to become an adult. Since adult fireflies only live a few weeks, they're clearly just the tip of the firefly iceberg.

Larvae of the common European glow-worm, *Lampyris noctiluca*, forage aboveground. They're commonly seen in gardens and meadows, by the roadside, and along railroad tracks. We know quite a bit about their habits because naturalists have studied these large and conspicuous larvae for more than a century. So it happens to be the glow-worm who appears as the main character in the following short play:

* * *

*Vagabond Stars: A Portrait of the Firefly as a Young Beetle*

Act I, Scene 1: The curtain opens on a faintly glowing firefly egg nestled into some moss, where its mother deposited it about three weeks ago. Since early July, it has faced the hazards of desiccation and egg predators all alone—and survived. Something inside the egg moves, and struggles to escape its shell. Now a tiny grub-like larva hatches out. With poor eyesight, this six-legged baby relies on its sense of smell to make its way in the world.

Act 1, Scene 2: Nighttime, and the larva is feeling very hungry. On foot, it sets off to hunt for every baby's favorite meal: a big, juicy snail! Crawling along the ground at a few meters per hour, the larva suddenly smells *eau de l'escargot* with its stubby little antennae. It follows the slime trail to its first victim: a handsome garden snail, which dwarfs the tiny larva. Undeterred, the larva climbs onto the snail's shell, then reaches over and probes the soft flesh with its mouthparts. On the video screen behind the stage, we see a close-up of the formidable weaponry the firefly larva will use to subdue its prey: it wields a pair of sickle-shaped jaws, each curving inward to a sharp point. Near the tip of each jaw, a small hole is barely visible, the opening of a narrow tube connected to the larva's midgut. Gently nipping at the snail, the larva uses these weapons to pierce the snail's skin and inject paralyzing toxins. Bitten, the snail tries to flee, but the firefly larva rides along, clinging tenaciously to the snail's shell. More bites, and the snail slows. Another, and the snail finally stops moving.

Act 1, Scene 3: Now the voracious larva settles in to consume its immobilized prey, whose beating heart reveals that it is still alive: its freshness is guaranteed! The larva begins to liquefy its victim by sinking its jaws into the snail's flesh and injecting it with digestive enzymes. For the next three days and nights, the glut-

**FIGURE 2.2** Baby fireflies are fierce predators, using their hollow jaws to inject immobilizing toxins and digestive enzymes into their prey. Here, a larva of the common European glow-worm attacks a snail (*Lampyris noctiluca* photo by Heinz Albers).

tonous larva feasts on snail soup. As the larva eats, its body visibly expands until it can no longer be contained within its exoskeleton. To accommodate its growing body, the larva must now discard its old skin and replace it with a larger one.

INTERMISSION

Act 2, Scene 1: Midsummer—the longest day approaches. This larva has taken its job seriously: over the past eighteen months it has consumed seventy snails, shed its skin several times, and increased its body weight three-hundred-fold. Yet for the last few weeks, the larva has been on walkabout. It's been wandering restlessly, seeking shelter suitable for a life-changing transformation. This vagabond now crawls beneath a fallen log where other wandering larvae have gathered. Curling up, each lies motionless for a few days, and then sheds its final larval skin. It is now a pupa. For two weeks all the pupae huddle together, hidden under their log. They hardly move, though when they're disturbed they wriggle and glow brightly. On the inside, they're hard at work completely dismantling their old bodies and painstakingly assembling new ones.

Act 2, Scene 2: Nighttime. Beneath the log, newly minted adults are struggling to shrug off their pupal casings. One by one, they crawl out and into their new lives. Some of these adults are large, plump, and wingless: these are female glow-worms. Others are only one-tenth this size, but they've got wings and are ready to fly: these are males. All these adults have lost their appetite for food; what's on their mind now is sex. Driven by the urge to procreate, during the final two weeks of their lives these adult fireflies must rely entirely on whatever reserves they've managed to accumulate during their months of larval feasting. After they disperse, they'll be spending this capital down to zero to fuel their new obsessions: courtship and mating, then fertilizing and laying eggs.

* * *

Ah, romance clearly lies ahead . . . but this enticing next act must wait until chapter 3. For now, let's return to *Photinus carolinus*, the captivatingly synchronous firefly of the Smokies, to see how it spends its grubby childhood before it emerges as a radiant adult.

Sometime in June, those *Photinus carolinus* females that have mated will deposit their eggs singly in moist soil or moss. If all goes well, about two weeks later these eggs will hatch into gray-bodied, tiny larvae about 2–3 millimeters long. For the next eighteen months these larvae will be hidden away underground, so we're unlikely to ever see one. They'll spend this time doing what all firefly larvae do best—eating voraciously and growing bigger.

These larvae happen to specialize in eating earthworms, a dietary mainstay for larvae of *Photinus* fireflies. Like glow-worm larvae, they're able to subdue prey many times their size. They bite the worm repeatedly with their sharp jaws, each time injecting neurotoxins. Sometimes *Photinus* larvae will gang up to attack an earthworm together. Once it's been immobilized, they'll feed gregariously on it for days.

In the course of my research I've raised quite a few *Photinus* larvae, keeping them happy by feeding them earthworms every week or so. Comically gluttonous, they often eat so much that their hugely distended abdomens prevent their legs from reaching the ground. They're forced to lie upon their backs, legs waving uselessly in the air, until they've digested enough that they're able to walk again. By now my family has seen this so often that whenever we happen to overindulge ourselves at a very good meal, one of us is likely to comment "I'm as stuffed as a firefly larva!"

*Photinus* larvae spend their first summer and fall hunting earthworms and growing larger. As winter sets in these underground larvae become dormant, then they start feeding again the following spring. If they get big enough, these larvae might pupate the next summer. But more likely they'll continue foraging all through their second summer and fall, and will hibernate again the following winter.

After yet another spring of feasting, one day in May these larvae will gather together in a patch of moist soil or under rotting logs. Each constructs a tiny, igloo-shaped soil chamber, which it will occupy for the duration of its pupal stage. A few weeks later they will finally emerge as adult fireflies—the stars of the

Smokies Light Show, filling the sky with their dazzling lights. But it's the firefly babies who star in the next story about how these bright lights first evolved many eons ago.

## Their Glow Means No

Firefly larvae glow from a pair of small spots on the underside of their abdomens, near the tip. They glow when they're disturbed by touch or vibrations nearby, and they often glow while they're crawling around. Their larval lights shine on throughout the pupal stage, but typically disappear once the adult emerges. Larval lights are universal among fireflies: every known firefly species can light up during its larval stage, even when its adults are lightless. When adults do produce light, it shines from an entirely new lantern that gets assembled during the final stages of metamorphosis.

**FIGURE 2.3** All fireflies glow during their childhood. Shown here in their early life stages are three firefly larvae (on the left) and two firefly pupae (photo by Siah St. Clair).

In 2001, Marc Branham and John Wenzel, biologists then at Ohio State University, made a surprising discovery using phylogenetic analysis, a tool that lets us peer backward in evolutionary time by retracing branches in the tree of life. Using morphology (the physical form) of present-day fireflies and related insects, these scientists found evidence that fireflies' light-producing talent first arose within the larval stage of an early firefly progenitor. But why would an ancient firefly larva need to make light? Larval insects aren't sexually active—they're much too young to be looking for love!

Many poisonous or distasteful animals use bright coloration—often yellow, orange, red, and black—to warn off potential predators. Most vertebrate predators are pretty smart; once they attack such prey, they quickly learn to associate the bright color pattern with a bad gustatory experience. From then on, they'll avoid future encounters. For instance, the distinctive orange-against-black colors of Monarch butterflies warn birds and other insect-eating predators that these butterflies are poisonous. When naive blue jays—who apparently don't know any better—eat a monarch butterfly, they'll quickly vomit it back up. And after one such distasteful experience, they tend to reject all other monarchs.

But firefly larvae are active mainly at night or underground, where such bright colors would be futile. A light in the darkness, on the other hand, would be quite noticeable. We also know that larval fireflies taste terrible. As chapter 7 reveals, many insect-eating predators like birds, toads, and mice show classic aversion reactions to fireflies—they wipe off their snouts or beaks, gag, and run away. So ample evidence suggests that fireflies' bioluminescence first evolved to help baby fireflies ward off predators: like a neon warning sign, it blazed out "I'm toxic—stay away!" Millions of years would elapse before these larval lights got co-opted to become a courtship signal for adult fireflies.

## Creative Improvisation: Fireflies Evolving

As I've traveled around the world, I've encountered many people who believe that God created fireflies for humans to marvel at and enjoy. I can certainly understand how fireflies inspire such reverence. Some nights, when I've been lucky enough to be standing out in a field surrounded on all sides by these silent sparks, I've felt a profound connection to the entire universe. Yet for me, the most miraculous thing about fireflies is how beautifully they illuminate the evolutionary forces that have shaped all life on Earth over the past 3.8 billion years.

Evolution is a complex, unchoreographed dance of creative improvisation. It starts off with something old, then—without intent or foresight—comes up with something new. When life takes that new something out for a spin, it might turn out to be something better. Or it might turn out to be something worse. Each new model gets different mileage, measured in evolutionary terms by how many offspring it can launch into the next generation. Over lengthy stretches of time certain new models come to replace the old ones, just because they happen to get better mileage. This process of random tinkering goes on and on for gigayears, gradually ratcheting the standards higher and higher by adding small improvements along the way.

Like all other living creatures, fireflies have been sculpted by two powerful forces: natural selection and sexual selection. In *The Origin of Species*, Charles Darwin described the pervasive evolutionary force he called natural selection. Because individuals have slightly different versions of inherited traits, they'll differ in how well they can garner necessary resources and avoid getting eaten by predators. As Darwin wrote in a wonderfully poetic passage:

> It may be said that natural selection is daily and hourly scrutinising, throughout the world, every variation, even the slightest; rejecting that which is bad, preserving and adding up all that is good; silently and insensibly working, whenever and wherever opportunity offers, at the improvement of each organic being in relation to its organic and inorganic conditions of life.

In Darwinian natural selection, this variation among individuals translates into stark, all or nothing outcomes: some will live, while others will die. It was natural selection, then, that first drove the evolution of fireflies' bright lights. Once juvenile fireflies struck upon a biochemical innovation that discouraged predators, this trait stuck around because it helped them survive their lengthy childhood.

Sexual selection is a much more subtle evolutionary force, because it's based on varying degrees of reproductive success. In his second-most-famous book, *The Descent of Man, and Selection in Relation to Sex*, Darwin painted a vivid, admiring picture of the bizarre and extravagant features sported by the males of so many creatures: the incessant croaking of male frogs, the unwieldy horns held aloft by male rhinoceros beetles, and the elaborate plumage proudly displayed by strutting peacocks. Surely these extravagant male ornaments and armaments couldn't have arisen through natural selection—after all, how could these features possibly help males escape from their predators or gather food?

Instead, Darwin argued, such male flamboyance must have evolved because it somehow improves the owner's reproductive prospects. Animals don't reproduce for the sake of their species. They reproduce for their own sake—well, actually to ensure that their own genes get transmitted to the next generation. And all the weaponry and frippery of the best-endowed males does indeed give them a mating advantage. Sexual selection operates through two distinct channels. Some traits work by enhancing a male's ability to outcompete his rivals. Other traits give their owner a reproductive leg up by making him irresistibly charming to the ladies.

Sexual selection was the evolutionary maestro behind the mesmerizing scene I'd witnessed that night in the Smokies. Out in the forest, I'd seen hundreds of male *Photinus carolinus* all trying to attract attention from some females perched down below. Their bioluminescent frenzy was the behavioral equivalent of a peacock's flamboyant tail. These lightningbug fireflies also offer a stunning example of evolution's creative improvisation. Long ago, some lightningbug ancestors had repurposed their larval warning lights into a highly versatile tool that would bring firefly courtship to a whole new level.

## Synchronous Symphonies

Among all the glamorous rituals shaped by sexual selection, perhaps the most spectacular are the displays performed by certain synchronously flashing fireflies. For reasons we don't yet understand, only a few lightningbug fireflies show this special ability to flash in unison with one another.

In Tennessee, the males of *Photinus carolinus* constitute a traveling symphony: while in flight, they join with nearby males to synchronize their six-pulsed courtship flashes. In Southeast Asia, meanwhile, certain male fireflies gather in stationary aggregations called leks and synchronize their flashes. One such stationary synchronizer is *Pteroptyx tener*, known in Malaysia as *kelip-kelip*. Each night, thousands of these males congregate in particular mangrove trees along tidal rivers. Perched singly on leaves, they put on a communal display by flashing all in unison. Relatively rare among insects, such leks are popular mating systems with certain birds like peacocks, birds of paradise, and sage grouse. The sole purpose of these aggregations seems to be for males to strut their stuff in front of discriminating females.

Fireflies' synchronous symphonies—whether traveling or stationary—are entirely silent. But to me, each display somehow conveys a unique musical quality. The six bright, brassy flashes of *Photinus carolinus* resemble trumpet blasts blaring out into the forest. *Kelip-kelip* fireflies seem more like an entire orchestral violin section playing pizzicato in unison. But, impressively, these fireflies achieve their synchrony without any conductor—no leader is calling out their rhythm.

We now have a pretty good mechanistic understanding of *how* such fireflies manage to synchronize their flashes, though I'll defer this discussion until chapter 6. But *why* do fireflies synchronize? Males should be competing for females' attention, so why should they cooperate to synchronize their courtship signals? Because their spectacle is seemingly so paradoxical, it's worth considering several proposed hypotheses for how synchronous courtship signals might have evolved.

One idea is that synchrony arises as an accidental by-product of competition between males to attract females' attention. In many creatures, including frogs, crickets, and cicadas, males gather to croak, chirp, and click out their audible courtship signals, and their rhythmic calls are often synchronized to within a few seconds of one another. Studies of these acoustic signalers show that females are attracted to the first signal they perceive when male calls are concentrated in time. If males can adjust their timing in response to one another, this bias in females' attention can lead to synchrony as each male tries to signal first to improve his own chances. In this case, synchrony itself provides no benefit to the participating males; it merely emerges as a by-product of females' perceptual bias. Evidence suggests that such "synchrony by default" provides a good explanation for the synchronous acoustic choruses of many insects. We don't know if this holds for fireflies: although females of the Big Dipper firefly, *Photinus pyralis*, do show this perceptual bias when tested, males of this species typically don't synchronize.

Alternatively, firefly flash synchrony might be a truly cooperative behavior that evolved because it provides some benefit for each participating male. Several suggestions have been made for just what these benefits might be. Three ideas focus on how synchrony might improve signal detection.

The first is called the "rhythm preservation" hypothesis: when males synchronize their signals, together they can clearly broadcast the flash rhythm that's characteristic of their own species. If several different firefly species happen to overlap in time and space, males who synchronize with one another could benefit by making it easier for females to recognize their own species' signal. Studies done by Andy Moiseff and Jon Copeland in Elkmont on the traveling synchronizer *Photinus carolinus* support this rhythm preservation idea. They used arrays of LEDs (light-emitting diodes) to see how females react to male-like flashes delivered either synchronously or asynchronously. When females saw several six-pulsed signals delivered synchronously, they responded more often than when they saw identical signals scattered in time. A female's visual field apparently gets cluttered when nonsynchronized signals come in from several males at once. Because males are constantly moving, females may have difficulty picking out the flash pattern given by any single male. So at high densities, a *Photinus carolinus* male might benefit from falling into step and flashing synchronously—this behavior encourages females to respond by making it easier for them to detect males of their own species. Unfortunately, similar tests haven't yet been done on females belonging to stationary synchronizers like the *Pteroptyx* fireflies of Southeast Asia.

A complementary notion: synchrony might provide a "silent window" that allows males to detect a female's response more readily. In many fireflies, the two sexes exchange flash signals in a back-and-forth dialogue. Females typically give their response flash during the brief dark period that falls between successive male signals. By synchronizing flashes, a firefly male might improve his ability to detect a female's response flash because it falls into everyone's silent window.

One final idea that's been proposed to explain why flash synchrony has evolved is called the "beacon" hypothesis. The notion here is that males broadcast a brighter signal when they cooperate to synchronize their flashes. In dense vegetation, such a collective effort may produce a signal that's visible to females at greater distances. Flying between display trees, females are likely to be attracted to the brightest beacon. Although still untested, this idea could help explain synchronous flashing in *Pteroptyx* fireflies, where stationary males gather in particular trees.

We don't yet have the necessary evidence to distinguish among these ideas, so exactly why fireflies synchronize remains an intriguing mystery. Each hypothesis explains how an individual male who synchronizes his flashes might gain better access to potential mates. Yet Darwinian sexual selection predicts that males should compete with one another for mating opportunities. Indeed, chapter 6 describes how, once a female firefly appears on the scene, such paradoxically cooperative male behavior gets abruptly replaced by cutthroat competition.

* * *

A firefly's life is a journey of astonishing transformation. When its life began, it was a crawling, earthbound creature that did nothing but eat, grow, and try to survive. A voracious predator, the baby firefly wielded fearsome jaws that first paralyzed, and then liquefied, its prey. Forged by natural selection in the struggle for survival, its larval lights evolved hand in hand with chemicals that repelled predators. The firefly's life was nearly at its end when it finally metamorphosed into an adult. Now entirely obsessed with sex, the firefly will no longer eat. Although its courtship rituals vary according to what branch of the family it belongs to, each adult will spend the remainder of its life trying to win a mate and launch its offspring into the next generation.

So now we're poised to head out into the night to join some typical New England lightningbugs in their quest for reproductive success. We'll visit these flying sparks in July, when their mating season is in full swing, to get an intimate, behind-the-scenes look at their luminous sex lives.

CHAPTER 3

* * *

# Splendors in the Grass

*All thoughts, all passions, all delights,*
*Whatever stirs this mortal frame,*
*All are but ministers of Love,*
*And feed his sacred flame.*
- Samuel Taylor Coleridge -

## Wild about Fireflies

As daylight drains from the summer sky, a cool breeze rustles through a hay-scented New England meadow. Relaxing in the long grass, even the keenest observer might miss the miniature army that's now awakening from its daytime slumber. One by one, tiny male fireflies are creeping upward along grass highways. They pause at each apex, ready to lift off like silent Black Hawks. But as they prepare for their nightly search missions, these firefly males aren't motivated by military conquest. Their quest? Genetic immortality. They're hell-bent on procreation, driven by an urgent need to propel their genes into the next firefly generation. These resolute males are destined to spend every night of their short adult lives valiantly broadcasting their luminous signals. Unfortunately, the odds are stacked against them as they head off into the night looking for love.

These are *Photinus* fireflies, the most common lightningbugs in North America. We happen to know about *Photinus* sex lives in intimate detail—more than we know about any other firefly in the world. This deep knowledge is mainly due to a few US scientists who've dedicated lifetimes of summer nights to deciphering firefly behavior. One of these firefly mavens is Jim Lloyd, an entomologist and professor emeritus at the University of Florida. Born in 1933, Lloyd grew up near the Mohawk Valley in New York State, where he spent his boyhood fishing, hunting, and just hanging around outdoors. Lloyd hated school. He did a stint in the US Navy, sold shoes, and stirred batter in a cracker factory, then decided to try out college. When a class project sent him out into a local marsh to observe fireflies, he loved it. He was also startled to discover how little was known about these curious insects. His newfound fascination lasted him through graduate school and beyond—actually, for a lifetime.

In the mid-1960s Lloyd earned his PhD from Cornell University by chasing *Photinus* fireflies through fields, forests, and marshlands. During these years, Lloyd spent several summers crisscrossing the United States in his pickup truck. Each night, he'd drive slowly along the back roads, headlights off, head stuck out the window looking for flashes. Whenever he spotted an active firefly hangout, he'd pull off the road and set up camp nearby. Armed with a stopwatch, a flash-recording device, and a battery-operated chart recorder, Lloyd would carefully note the behaviors and flash styles of whatever fireflies he could find for the next few nights. He also collected a few specimens corresponding to each flash style, later examining their microscopic features in order to pin down what species they belonged to. A self-described hermit—Lloyd still boasts of his "legendary asocial nature"—he loved the solitary lifestyle that accompanied this intensive fieldwork. By the time Lloyd finished his dissertation, he'd managed to decipher the secret codes that fireflies use to communicate with their prospective mates.

Lloyd landed a job at the University of Florida, where he poured his firefly passion into a perennially popular honors course called Biology and Natural History with Fireflies. Never a big fan of lectures, he instead handed out informal discourses dredged from his deep knowledge of firefly natural history, spiced with questions designed to pique his students' interests. Students carried their curiosity out into the night on field trips to study firefly mating rituals, predation by spiders, and larval feeding preferences.

Even though he's long since retired, everyone still refers to Lloyd as "The Firefly Doc." And this moniker fits him perfectly. Over fifty+ years, Lloyd has totaled more than three thousand field nights out in the wild, obsessively cataloging how North American fireflies behave in their natural habitat. His entire packet of publications—more than one hundred scientific papers along with a dozen or so book

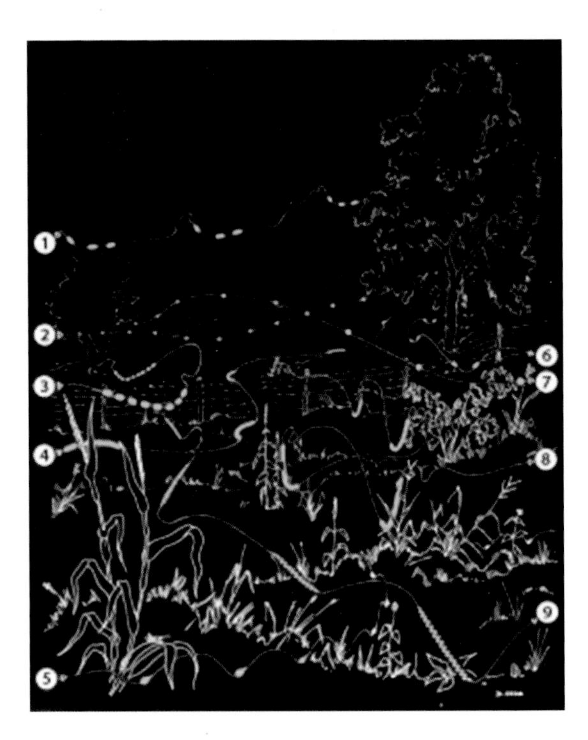


**FIGURE 3.1** Males from nine different species of North American *Photinus* fireflies show distinctive flight paths and flash patterns (illustration by Dan Otte, from Lloyd 1966).

chapters—is dedicated to understanding firefly behaviors and how they might have evolved.

I first met this quiet, intense, and immensely knowledgeable man when he accepted my invitation to give a lecture at Duke University while I was there studying for my PhD degree. In spite of a cantankerous reputation and his apparent discomfort at wearing a jacket and tie, Lloyd had a gentle manner. Delivering the most entertaining academic talk I've ever heard, Lloyd vividly illustrated what he'd deciphered during his field studies. In the darkened auditorium he took out a homemade contraption that looked like a fishing pole, except that in place of a hook dangled a tiny light. Using the finger trigger of this device, Lloyd quite memorably cavorted across the stage to illustrate the distinctive flash patterns used by different *Photinus* firefly species.

North America has about thirty-five different *Photinus* species. In the 1960s Lloyd discovered that males of each species emit a different pattern of flashes (Figure 3.1). Depending what species he belongs to, a *Photinus* male will court by emitting a flash pattern with one, two, or several pulses of light. After a short pause, he repeats the whole pattern again. Each flash pattern is distinctive in how many light pulses it contains, how long each pulse lasts, and the dark interval between pulses.

So it's the *timing* of a male's flash, rather than its color or shape, that conveys the crucial information about his species identity and sex. Like mariners who use the distinctive light pattern to tell which lighthouse they're approaching, female fireflies use such differences in flash timing to distinguish the males of their own species. Later in this book, I'll explain how you too can talk to fireflies using Lloyd's discoveries about firefly linguistics.

## Defining the Indefinable

Not only did Lloyd decode the language of love for *Photinus* fireflies, but he also discovered several entirely new firefly species. Yet how did he know when he'd found a new species? And what *is* a species? It's a common household word, conveniently both singular and plural. We count species to measure biodiversity. And we decorate museum specimens with species labels, formally inducting each carefully stuffed bird, pressed seaweed, or pinned insect into our human taxonomy. Yet it's proven surprisingly slippery for biologists to define exactly what they mean by a species. Sex forms the bedrock of one widely accepted definition, called the biological species concept. In it, two groups are considered to be the same species if they can successfully mate with each other and produce fertile offspring. Reproductively united, a species is defined by its members swimming around together in a common gene pool. When two groups are reproductively isolated—each swimming in its own private little genetic puddle—then we refer to them as different species.

Lloyd decided he'd shine the light of this biological species criterion directly onto some of his *Photinus* fireflies, out in the wild. He wanted to see if they were in truth reproductively isolated from one another. In one site-exchange experiment, he collected six responsive females from a Branchville, New Jersey, population of *Photinus scintillans* and drove them over to Silver Springs, Maryland. The next night he positioned these females, each in her own glass jar, in a breeding hotspot for *Photinus marginellus* fireflies. Would the *scintillans* females answer the ardently courting *marginellus* males flying around their jars? He also transported some *marginellus* females back to the *scintillans* site to see how they responded to the other males' courtship flashes. Lloyd tenaciously performed these site-exchange experiments for thirteen pairs of putative species. In nearly every case, he discovered that females remained true to their own gene pool—they only flashed back to their own males' courtship signals. Lloyd's results showed that *Photinus* admirably upheld the biological species criterion—they were distinct species because they didn't interbreed.

Back from the field, Lloyd started visiting natural history museums where he could compare his fireflies to the taxonomic reference specimens housed in their collections. The formal scientific descriptions of insect species are generally based on dead bodies, and so by necessity they rely on anatomical similarity to distinguish one species from the next. Lloyd noticed that some of the *Photinus* fireflies he'd collected were anatomically identical, right down to the voluptuous curves on their microscopic genitalia. So according to the formal descriptions, they should all have belonged to a single species. But Lloyd had seen these insects alive, flashing their hearts out in the wild. He'd seen them using completely different courtship signals. And based on these different courtship signals, they didn't interbreed even when he gave them ample opportunity. Even though they were indistinguishable once dead, these living fireflies clearly had no trouble telling each other apart. So Lloyd proceeded to write up new scientific descriptions using differences in flash behavior to distinguish these once clandestine species.

Defining a species can be even more slippery in other creatures, including *Photuris* fireflies in North America. *Photuris* fireflies can quickly switch between different flash patterns, so their flash behavior is not as useful in categorizing different species. And sometimes what *look* like two distinct species can actually interbreed to produce intermediate hybrids. Even Charles Darwin, a man who spent a lot of time pondering species, was not too enthusiastic about defining them. "It is really laughable to see what different ideas are prominent in various naturalists' minds, when they speak of 'species,'" Darwin wrote in a letter. "It all comes, I believe, from trying to define the indefinable." While the concept of biological species is undeniably practical, such categories often have fuzzy boundaries.

## Heading Out into the Night

Let's head back to our New England meadow and rejoin these tiny insects sparking through the night in search of love. This meadow is home to a firefly species known as *Photinus greeni*, whose males give a distinctive pair of quick light pulses separated by 1.2 seconds. Poised atop grass blades, our army of firefly males has waited patiently for darkness. Now each male lifts off into the air to start his nightly patrol. As he flies, he earnestly advertises his availability about every 4 seconds by flashing out to anyone who's watching: "I'm a *greeni* male, here I am! I'm a *greeni* male, here I am!" Each time he shines his light, the male pauses for an instant in hopes of spotting a female. So tonight across the meadow, it's *wink*, *wink*, hover and hope . . . *wink*, *wink*, hover and hope . . . *wink*, *wink*, hover and hope. Males concentrate their efforts in places where females are likely to gather and then fly along to the next likely pickup spot. Hundreds of flying males now fill the meadow, their flashes scintillating like sunlight glinting off the sea.

What's a firefly's worst nightmare? Sometimes I think it must be the sudden rainstorm that hits just when they're starting their courtship flights. One night I watch as raindrops slam into their tiny bodies and drive them down to earth like sodden shooting stars. Wet wings will mean no more flying for tonight. These soggy males must now continue their courtship on foot, flashing sporadically as they plod tediously through the grass in their search for females.

But where are those obscure objects of their desire, the females? Even though female *Photinus greeni* can fly, they rarely waste their precious energy doing so. Instead, they perch quietly on grass blades, like women perching on barstools at a singles bar. If a female spots an especially attractive male, she might deign to acknowledge his advances: she'll flash him a come-hither reply, aiming her lantern in his direction. *Photinus* females usually give a single prolonged flash that rises in a crescendo and then slowly fades out. Here again, timing is everything. Across different *Photinus* species, females differ in how long they wait before flashing out their response. And males use these different response delays to identify females belonging to their own species. In *Photinus greeni*, females have a very short response delay: after the male flashes, they wait less than a second before offering their crescendo response.

## A Light Snack

While our firefly male is flying and flashing, his mind set on sex, other nocturnal creatures are out here, too—and they're intent upon dinner. This meadow is home to hundreds of wolf spiders and their orb-weaving cousins, all eager to put firefly filet on tonight's menu. They've strung webs between the tallest grass stalks, invisible traps waiting to ensnare careless flyers. Lots of unlucky male fire-

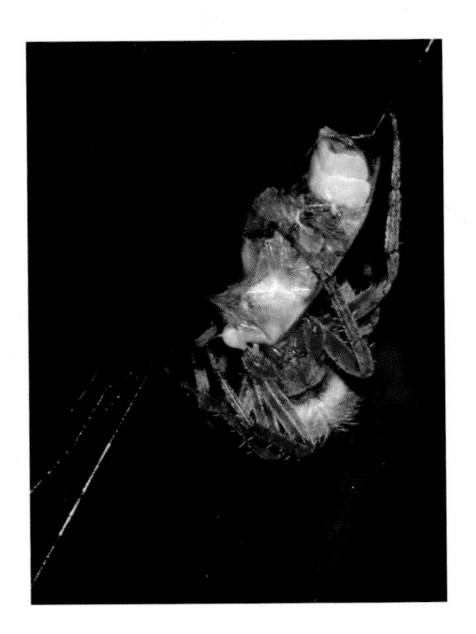

**FIGURE 3.2** A hapless male whose courting days ended abruptly when he got trapped and wrapped by an orb-weaving spider (photo by Van Truan).

flies will get quietly filtered out of the air tonight, and so will meet their sticky ends wrapped up tightly and hanging limply (Figure 3.2). Yet even as they die, the unfortunate male's flash may live on. Our field notebooks are filled with observations of fireflies shrouded in spider's silk that continue their rhythmic flashing long after they've been immobilized. Their flashing attracts other fireflies, and sometimes these newcomers wind up trapped in the web too. Perhaps these spiders have figured out how to turn their captives into a bioluminescent fishing lure.

What are the odds in this gamble between sex and death? Jim Lloyd decided to find out. He headed out from Gainesville, Florida, to a grassland hotspot for *Photinus collustrans* fireflies. Equipped with his sharp eyes, a surveyor's wheel, and a tally counter, Lloyd spent several nights doggedly tracking 199 males—one at a time—as they cumulatively logged more than 10 frequent flyer miles and gave nearly 8,000 flashes. Only two males wound up finding females, while the same number found themselves in a predator's deadly embrace. For a *Photinus* male, the search for mates is certainly a high-stakes game of reproductive roulette.

## Closer Encounters

Meanwhile, our male fireflies are still aloft back in their New England meadow. They've been patiently searching for nearly twenty minutes. Now, at last, one male glimpses a flashed reply from below. The time delay is just right for his species—finally, a female has answered his calls! He swiftly drops from the air to land nearby. A flirtatious flash exchange strikes up as he scrambles toward the female on foot. He runs partway up a grass blade, stops, and flashes. Will she answer him this time? No. He runs farther up the blade, stops, and flashes once more. Now she does reply, but rats—it looks like he's been traveling in the wrong direction. He'll need to retrace his steps. This intermittent courtship dialogue continues while the male runs frenetically up and down grass blades searching for the female. Another hour passes before he sees her response directly above him—she's right here on the same grass blade! Rushing upward, he climbs onto her back. Mating begins without delay as he couples his genitalia with hers. But before this relationship can be consummated, he'll need to swivel around to face in the opposite direction. Precariously perched on a narrow grass blade, the male needs considerable acrobatic skill to pull off this feat without tumbling to the ground. Once the pair has settled comfortably into a tail-to-tail position, their flashing ceases (Figure 3.3).

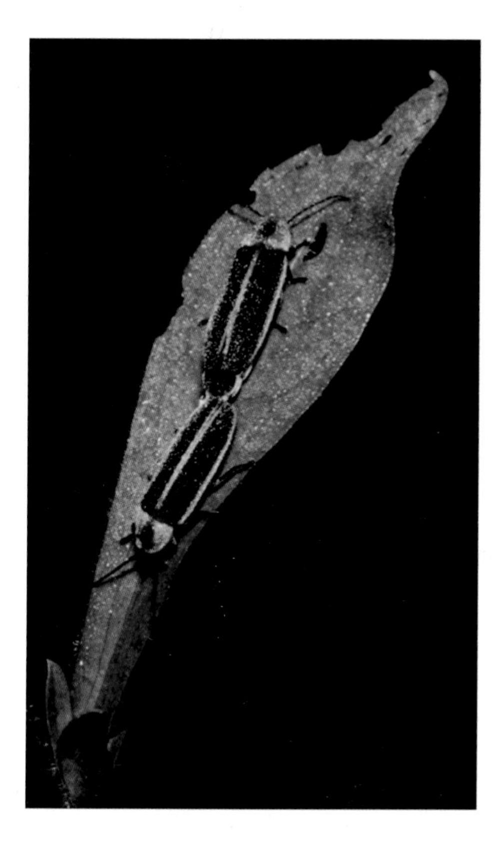

**FIGURE 3.3** A mating pair of *Photinus* fireflies (female above, male below; photo by the author).

But wait . . . what happens after the lights go out? When I first began researching fireflies in the 1980s, most studies had focused on the flashier side of firefly courtship. Once the actual mating began, most sane scientists had packed up their field gear and headed home to their cozy beds. But I was fascinated by firefly sex lives. How long did firefly pairs remain coupled? Were these only brief flings? Were fireflies speed daters, breaking up only to take another mating partner later that night?

To answer these questions, my students and I have spent many sleepless and mosquito-filled nights tracking *Photinus* fireflies in fields around Boston. Each night we arrived at the study site promptly by 8:35 p.m., with our clipboards ready and our headlamps equipped with blue filters—fireflies don't see blue very well. When the male flight period started at nine, we dashed about to find as many responsive females as we possibly could. After flagging the location of each female's perch, we marked her with a tiny dot of nontoxic paint. Then we sat down on our campstools to watch as each of our females started up dialogues with passing males. And we waited. And waited—sometimes firefly courtship conversations go on for hours. When a pair finally made contact, we recorded what time the copulation began. Then we started making our rounds. All night long, every half hour we checked to see whether each pair was still mating. When the sky lightened and the birds' dawn chorus started up, we were still dutifully scribbling "mating in progress" onto our data sheets. Not until dawn did these firefly duos finally uncouple, clambering down their matrimonial grass blades to go their separate ways. In spite of our sleep deprivation, we were excited to have discovered that *Photinus* fireflies mate only once each night. Their protracted liaison is what scientists call "copulatory mate guarding," a behavior male fireflies use to guard against females running off and hooking up with some rival later that night.

And so our firefly sweethearts have remained locked together throughout the night, reluctantly parting at dawn. It's all very romantic, yet can our firefly love story really be so simple?

## To the Victors Go the Spoils

No, it can't. In early 1980s the field of animal behavior was in flux. The nascent field of behavioral ecology was springing up to replace classical ethology, which had focused on describing fixed, invariant patterns of behavior. Instead, behavioral ecologists chose a more evolutionary approach: they wanted to find out how behaviors vary among individuals, and how such variation influences an individual's ability to survive and successfully reproduce. Jim Lloyd became one of the earliest advocates for applying Darwin's ideas about sexual selection to insects in general, and to fireflies in particular. As he collected his detailed natural history observations, he was constantly pondering how a firefly's form and behavior might have been shaped by sexual selection. Inspired by Lloyd's ideas, I began my own backyard explorations. These eventually turned into a decades-long research program that would reveal many intimate secrets about firefly sex lives.

I still recall the sweltering summer evening when I stumbled upon one startling firefly fact. I was sitting on my back porch in Durham, North Carolina, with my dog, a black lab named Orpheus. We were both enjoying the clouds of fireflies that rose up from the grass when I began to wonder whether males and females were equally represented within this horde. I grabbed my insect net (biology graduate students generally keep such things handy), jumped off the porch, and started sweeping up all the flying creatures I could gather. In short order I'd netted several hundred fireflies that I released into some large plastic jugs. Over the next few hours, I gently extracted them one by one, examining their lantern shape to tell whether I held male or a female in my hand. I was astounded when I realized I'd inspected over two hundred fireflies, but I had yet to see a single female.

A trip to the university library convinced me that I would need to search down in the grass to find these *Photinus* females, so that's what I did for the next few evenings. Orpheus, descended from a long line of hunting champions, became my first field assistant. We stayed up long past midnight, locating females by their response flashes. Each time I spotted one, I marked her location with a plastic survey flag. Once Orpheus figured out what I was doing, he put his pointing skills to good use. Looking over, I'd see him standing—his nose outstretched, paw up—telling me that he'd found another female firefly. When I counted up these flags, I was astonished to find we'd flagged only twelve females. This dating scene for male fireflies was ridiculously competitive: 218 males flashing their hearts out to only twelve females! During those few nights I gained a lasting respect for the long reproductive odds that male fireflies play against.

For nearly all organisms, the very *concept* of male and female is based on a fundamental asymmetry in their reproductive investment. It all starts with their gametes. By definition, females are the sex that produces eggs, those large, immobile cells chock full of organelles and other juicy cytoplasmic bits. Meanwhile, male sperm are little more than tiny mobile bits of DNA. Beyond this gametic asymmetry females generally invest more in caring for offspring before, and sometimes after, birth. In the 1970s, biologist Robert Trivers proposed that this fundamental sex difference in parental investment was ultimately responsible for the evolution of the classic courtship behaviors we see repeated so often throughout the animal kingdom. Males can generally be counted on to compete over females—some Australian beetle males are so eager to mate that they often die attempting to copulate with discarded beer bottles! Females, on the other hand, are coy and often very choosy about their mates. This nearly universal pattern, then, arises from differences between males and females in the relative investment each sex makes in producing offspring. Males—members of the lower-investing sex—are destined for a life of competition, while females—the higher-investing sex—have more at stake and are more picky. A corollary to Trivers's parental investment idea is that males should take on the more active and costly side of courtship. This includes not only expensive male ornamentation and weaponry, but also male behaviors like searching out females or showing off with daredevil displays.

It turns out that the male-biased sex ratios that so startled me in my North Carolina backyard show up again and again in insects. Male mate competition has led to the evolution of many extraordinary mating behaviors. For one thing, males often get a jump-start on metamorphosis, and turn into adults sooner than their corresponding females. This pattern of early male emergence is known as protandry, and it's common among butterflies, mayflies, mosquitoes, and fireflies. Male competition even compels some insect males to take child brides. In certain mosquitoes, spiders, and butterflies, males will jealously stand guard over an immature female, chasing off rival males and waiting patiently until she becomes sexually mature. Males of some tropical *Heliconius* butterflies will guard an immature female pupa for days, pushing their genitalia right through the cocoon so they can copulate just as soon as the female emerges. Some male fireflies also use this child-bride tactic, guarding immature females and then mating when the female crawls out.

Mate competition isn't just about being the first one to reach the female. The classic animated film *Bambi* features a highly charged scene in which two adolescent male deer lock antlers over their common sweetheart. In many beetles, reptiles, and mammals, males have evolved horns, antlers, tusks, and spurs that they wield in fierce battles against rival males. In fireflies, though, such male-male interactions are a lot subtler. Males that are lucky enough to spot a female and engage her in a flash dialogue rarely have a chance to talk privately. Like a magnet, flash dialogues quickly attract other males. Very soon, each female finds herself surrounded by a small coterie of suitors, all flash-competing to capture her attention.

If you were to set up a campstool and spend some time carefully watching these competitive courtships, you might get to see some especially sneaky male behavior. When several males are all courting a single female, sometimes one male will slyly give a pseudo-female flash in response to a rival male. Looking just like a female flash, the male's pseudo-female flash crescendos, then slowly fades out; he also flashes using the female's typical response delay. Based on my own experience trying to locate females, I know these transsexual flashes can be pretty convincing. It appears these males have figured out a clever way to decoy competitors away from the actual female.

Even though they lack overt weaponry, competition among male fireflies sometimes gets physical. In *Photinus pyralis*, the Big Dipper firefly, sometimes as many as twenty rival males will surround a single female, creating a writhing love knot. Males vigorously push and shove each other, using their head shields to dislodge their rivals. Eventually one male wins out and mates with the female, although what exactly makes a male victorious in these battles isn't yet known. Poor losers or ever hopeful, several of the losing males often pile on top of the happy couple, stacking themselves up to six deep. Perhaps this is why male *Pteroptyx* ("bent-winged") fireflies from Southeast Asia use their wing covers to tightly clamp around their mate's abdomen—it seems a good way to prevent hostile takeovers.

Seen through the lens of sexual selection, we now have an entirely different view of the lovely luminous display out here in this meadow. It stars hundreds of male fireflies flying and flashing their hearts out in a quest for genetic survival. Male fireflies spend night after night flash-competing with scores of rivals to locate scarce females. These nocturnal flights require a lot of energy and are fueled only by whatever the male has managed to store up during his larval feasting. And as we've seen, flight isn't the only courtship expense incurred by males; their search missions are also fraught with dangerous predators. In fireflies, courtship is definitely a big-ticket affair, and the cost rests squarely on the males' shoulders.

## Ladies' Choice

Considerably more controversy would engulf the concept of female choice, Darwin's second mechanism of sexual selection. In his 1871 book, Darwin worked his way methodically through the animal kingdom giving minutely detailed examples of extravagant male ornaments that show up in creatures from crustaceans to insects, to fish, amphibians, and reptiles. He lingered lovingly for several chapters as he described sexual ornaments among birds:

> Male birds . . . charm the females by vocal or instrumental music of the most varied kinds. They are ornamented by all sorts of combs, wattles, protuberances, horns, air-distended sacs, topknots, naked shafts, plumes and lengthened feathers gracefully springing from all parts of the body. The beak and naked skin about the head, and the feathers are often gorgeously coloured. The males sometimes pay their court by dancing, or by fantastic antics performed either on the ground or in the air. In one instance, at least, the male emits a musky odour which we may suppose serves to charm or excite the female.

How did such varied and widespread male sexual ornaments evolve? Darwin proposed that females must somehow use these features to evaluate and choose their mates. He proceeded to marshal hundreds of examples to support his contention that females judge males' courtship displays, vocalizations, plumage, and other ornamentation, choosing to mate only with those males they find particularly "beautiful." Always quick to anticipate possible objections, Darwin went on: "No doubt this implies powers of discrimination and taste on the part of the female which will at first appear extremely improbable."

And he was right. Male scientists in the late nineteenth century, including Darwin's colleague Alfred Russel Wallace, were convinced that choosing mates required cognitive powers far beyond the capabilities of female animals. While Victorian England readily accepted the idea that males would fiercely compete for females, it ran against the cultural grain to claim that females, especially human ones, might also play an active role. During the early decades of the twentieth century, evolutionary biologists were caught up in the Modern Synthesis, which knit together Darwinian selection with Mendel's discoveries about genetic inheritance. Their central focus was natural selection and the role of mutation in creating genetic variants, the raw material for selection. For decades, Darwin's idea of sexual selection through female choice languished, ignored and all but forgotten by the scientific establishment.

It was not until the mid-twentieth century that Darwin's idea of females actively choosing their mates began to attract rigorous scientific scrutiny. Two parallel avenues of research would finally elevate female choice into a widely accepted mode of sexual selection. During the 1930s, laboratory studies with *Anolis* lizards showed that females paid close attention to males' bright red dewlaps as well as their vigorous push-ups and head-bobbing displays. At the same time, the population geneticist and statistician Sir Ronald Fisher published his book, *The Genetical Theory of Natural Selection*. In it, Fisher offered a theoretical explanation of how female choice for some arbitrary male trait could trigger this trait to get rapidly exaggerated over evolutionary time. The field exploded, and sexual selection researchers focused on female choice for the next several decades. By the 1990s, towering stacks of scientific studies had established that females *do* actively choose their mates, and that such decisions are based on small differences in male appearance and behavior.

Through years spent in the field watching close-up courtship interactions of fireflies, I'd noticed that female *Photinus* are quite picky. Even the most ardent suitor is rarely favored with a reply: *Photinus* females typically answer fewer than half of the male courtship flashes they see on a given night. When a female likes a particular suitor, she'll show this by responding more reliably to his flashes. And whichever male can elicit the highest rate of female responses is usually the one who gets the girl. For a firefly male, then, merely getting a female to reply will give him a major leg up in the quest for reproductive victory.

So what exactly does a female firefly consider sexy?

Over the past fifteen years, firefly biologists have come up with some pretty clever experiments to help answer this question. Many years back, Jim Lloyd's work had shown that *Photinus* females use male flash timing to distinguish between males belonging to different species. But even when male fireflies all belong to the same species, their flashes can be subtly different. This male-to-male variation is often too slight to be detected by the human eye, but it's easy to detect when you record and examine different males' flashes using a computer. Within some species, the males have different flash *durations*, while other species' males differ in their flash *rates*. And female fireflies appreciate these flash differences, too.

When it comes to deciphering exactly what females pay attention to during male courtship displays, animal behaviorists often use playback experiments. For instance, to figure out what male songs are preferred by female crickets, frogs, or birds, researchers use loudspeakers to play back different songs. They can then see which song variant attracts the most females. Similar playback experiments have been conducted to find out what kind of flashes female fireflies like best. It's relatively easy to mimic the flashes given by male fireflies by hooking up light-emitting diodes (LEDs) to a computer. When a female sees a flash she considers attractive, she'll give a flash response—how convenient! This photic playback technique has been used to poll females from many different firefly species on the all-important question: what's the most attractive flash of them all?

Firefly researchers Marc Branham and Mike Greenfield, then at the University of Kansas, first used this trick in 1996 to find out what female *Photinus consimilis* fireflies like best. Males in this species court females using a flash pattern consisting of six to nine pulses, the whole pattern being repeated at intervals. These researchers visited Roaring River State Park in Missouri, where they video-recorded the flash patterns given by sixty-one males that were searching for females. When they analyzed these recordings, they discovered these firefly males vary considerably in their flash signals. The researchers also brought some females back to their laboratory and played for each female a series of carefully engineered flashes. They separately changed how many pulses were in each flash pattern, the duration of each pulse, and the pulse rate, while they held both flash color and intensity constant. By keeping track of whether or not each female responded to each flash variant, they discovered that females didn't care much about the exact number or duration of pulses that that they see. But *Photinus consimilis* females really sat up and paid attention when the researchers played different pulse rates for them. Every single test female eagerly responded whenever she saw flash patterns with faster pulse rates, yet she completely ignored those with slower pulse rates. This was a pathbreaking experiment. First, it provided evidence that even *within* a species, males show subtle differences in their flash timing. Second, it revealed that female fireflies pay very close attention to differences among their prospective mates. Finally, it showed that sexual selection favors males that have faster pulse rates, because that's what females like.

Similar photic playback experiments have since been done with other *Photinus* fireflies, and these laboratory experiments reveal that for a firefly male, having the right flash is the key to capturing a female's heart. As part of his PhD research at Tufts, my graduate student Chris Cratsley worked with two other *Photinus* fireflies, both species whose males give single-pulsed courtship flashes. Chris discovered natural variation among *Photinus ignitus* males in their pulse duration, which varies between about 1/20 to 1/10 of a second. By wooing *Photinus ignitus* females with different flashes, Chris found they prefer longer pulse durations. *Photinus pyralis* fireflies also show similar among-male variation and female preference for longer pulse durations. So it seems that for *Photinus* fireflies, a male's flash timing not only conveys information about his species identity and gender, but also determines how much he appeals to the ladies. Firefly females apparently prefer more conspicuous male courtship signals, including those with faster pulse rates or longer pulse durations. This raises a perplexing question: based on such female preferences, why don't firefly males evolve even faster, or even longer, flash signals? As we'll see later on, other creatures with less amorous intentions are also keeping their eye on male courtship signals.

## Trading Places: Sex-Reversed Courtship Roles

We've seen that the classic courtship roles Darwin described—males compete while females are choosy—explain a great deal of firefly behavior. But sometimes the tables get turned. Males emerge as adults earlier than females, and they're also more likely to get eaten by predators. So by late summer there's often an oversupply of females. The ladies are still out there looking for action, yet potential mates are scarce. At this point, a female firefly can no longer afford to be choosy—instead she'll respond to nearly anything that flashes.

In fact, she'll respond to you! To find these late-season females, try walking through a firefly field a week or so after the male display has peaked. Take along a penlight and flash it toward the ground. You'll get answers from a chorus of females responding from their perches. This is one of my favorite firefly tricks—I always get a thrill seeing so many normally hard-to-find females lighting up simultaneously to answer my flash.

Late in the season, the few male fireflies still hanging around are in the groove because now *they* can choose among all these ladies. But why should males be picky? Because every male is always trying to father more offspring. In the mid-1990s my team of Tufts students discovered that these late-season firefly males actively choose the females who carry the most eggs. How can they tell? After mounting a female, a male uses his legs to gauge her girth. He's looking for the shapeliest female—she's the one with the biggest abdomen, because she'll have many eggs ready to be fertilized. Even though it takes extra effort for a male to court, find, and wrap his little legs around many different females before choosing the plumpest, the male firefly seems to find this an evolutionarily worthwhile strategy.

So concludes our nocturnal excursion into the mating habits of *Photinus* fireflies. For the male *Photinus greeni* fireflies we've been trailing, it's been an exhausting night spent in ardent courtships and passionate flash exchanges. But even for those few males who've successfully tied the knot, it's not over yet. Darwin's version of sexual selection was focused exclusively on mating success, but simply mating won't be enough for our firefly males. Our lucky male's reproductive quest must continue beyond tonight's successful mating, because tomorrow night his female might find another partner. So to win the ultimate evolutionary jackpot, this male needs to make certain he's the guy who will father most of her kids. To accomplish this, our firefly male will have to rely on some very different talents, as revealed in the next chapter.

Whenever I spend a night out in the field watching fireflies, I'm reminded of the man whose efforts provided the key to deciphering these silent conversations: Jim Lloyd—certified expert in firefly linguistics. One hot summer evening a few years back, Lloyd pulled up in his dusty old camper at my study site outside Boston. He was clearly ecstatic as we spent several hours tromping through muddy fields and tall grasses together, watching fireflies with a wonder that never grows old. Although Lloyd kept up a convincingly cantankerous front, I could easily see that nothing would ever diminish his delight at being outside in a field at night, surrounded by fireflies. At this he was, quite simply, genius.

CHAPTER 4

* * *

# With This Bling, I Thee Wed

*The great art of giving*
*consists of this:*
*the gift should cost very little*
*and yet be greatly coveted,*
*so that it may be the more highly appreciated.*
- Baltasar Gracián -

## After the Lights Go Out

People often ask how I first got started with fireflies. Surprisingly, I don't have many childhood recollections of chasing or collecting fireflies. It wasn't until I started my postdoctoral fellowship at Harvard that I began to really think seriously about studying fireflies. Why fireflies? Their visible, easy-to-decipher courtship signals and brief adult lives were a perfect match for my growing interest in the evolutionary game of sexual selection. As I've expanded my firefly studies over the past thirty years at Tufts University, I've been lucky to have many talented and hard-working students join my research team. Armed with their enthusiasm, insight, and curiosity, these bright young scientists-in-training have endured violent thunderstorms, voracious mosquitoes, and skunks, ticks, and poison ivy—all for the sake of better understanding firefly sex and evolution.

One especially hectic summer my husband began his clinical cardiology fellowship, so he was busy at the hospital most nights. That's when our five-month-old son became the youngest member of my research team. I carried Ben out to the field, carefully swathed in mosquito netting, and settled his infant seat down into the dew-covered grass. Every night, fireflies and stars whirled above him while my students and I gathered data about firefly sex lives. Sometimes I wonder if all those nights spent sleeping under the stars propelled Ben toward his career as a theoretical physicist, where he now gets to explore the mysteries of our universe.

Some of our most exciting firefly discoveries sprang from my scientific decision in the late 1980s to see what really happens after the lights go out. As described in the previous chapter, our work had already shown that female *Photinus* fireflies mated only once each night. But would they take another partner the next night?

To chart the mating history of individual fireflies, we used a low-tech but effective method. We flagged the location of each female we found and then carefully marked her wing covers with a unique pattern of tiny paint dots. Because they don't move around much, we could easily relocate these females after releasing them. We then went back night after night after night to find out what our individually marked females were up to. We recorded whether and when they responded to males' courtship signals, whether and when they mated, and with whom. For the entire summer we dutifully recorded every nightly firefly tryst.

Although they mated only once each night, it turned out that both sexes took many different mates over their two-week adult lives. I understand how this might

**FIGURE 4.1** The author searches for a female firefly who's been individually marked, aiming to track the firefly's mating history during every night of her short adult life (photo by Dan Perlman).

seem like an esoteric bit of knowledge. But while gallivanting males were no surprise, the discovery that firefly females had multiple mating partners had huge implications.

Studies of animal mating systems during the late 1980s were detecting an unexpected trend. Using sophisticated genetic paternity tests, these studies revealed that female promiscuity was rampant. This was true nearly everywhere biologists looked in the natural world: from dung flies to elk, from ants to ground squirrels, from field mice to fairy wrens, from bees to barn swallows, and, apparently, in fireflies. In every one of these creatures, the females commonly mated with—and produced offspring with—several male partners. Biologists were especially stunned by how sexually promiscuous the vast majority of birds turned out to be. In my college ethology course, birds had been cited as exemplary monogamists. Evidence of their close-knit family life surrounds us as pairs sweeten the air with their duets, cooperatively build nests, and diligently feed chicks together. Yet behind the scenes, female birds were definitely playing around. And biologists discovered these so-called extra-pair copulations weren't just for fun. Take the superb fairy wren (*Malurus cyaneus*), a charming Australian songbird. Male and female fairy wrens establish life-long pair bonds, working harmoniously with a single partner to raise their young. But fully two-thirds of the baby fairy wrens chirping in any nest belong to dads who are not their mother's main squeeze.

What difference does all this make? Quite a bit, as it turns out. Widespread female promiscuity challenged everything we thought we knew about sexual selection. Darwin believed that as long as a male managed to mate with a female, his evolutionary success was ensured. But now it was becoming clear that the tendrils of this key evolutionary process extended far beyond mere copulation. When each female mates with several partners, new innings get added to the reproductive game. Mating becomes necessary yet not sufficient, because these mating males won't share equally in fertilizing the female's eggs. Instead, they need to compete with one another to sire offspring. And even after they've chosen their mates, some females exert control over which males are successful in fathering their offspring. Discovery of female infidelity opened an exciting new frontier known as postcopulatory sexual selection. Over the past two decades, behavioral ecologists have unearthed some surprising strategies that animals use during and after mating to win this ongoing reproductive game.

## Sperm Wars, Sperm Love

So what's a male to do? He needs to win the ultimate evolutionary jackpot: to sire more offspring than any other males. Obviously he'll try to mate with lots of females. But he'll also need to battle for paternity with rival males to ensure that *his* sperm get priority treatment when it comes to fertilizing the female's eggs.

Driven by fierce sperm competition, males have evolved what might just be the craziest behaviors and weirdest structures ever seen in the animal kingdom. Male weapons like elk antlers and beetle horns have been forged by sexual selection for use in open battle. But sperm competition has fashioned more subtle weapons from animal genitalia, and these are wielded in secret battles fought deep within the females' reproductive system. Most female insects have special receptacles for storing the sperm they receive during mating. Sperm can survive for weeks, or even months, in these receptacles before a female uses it to fertilize her eggs. So whenever a male mates, he must not only transfer his own gametes, but also supplant any rivals' sperm that's already been deposited in the female's sperm bank. This challenge explains why male genitalia often show what Jim Lloyd described as "a veritable Swiss Army knife of gadgetry."

Insect penises carry an astounding variety of scoops, bristles, protuberances, and spines. Like the long wiry tools called plumbers' snakes, these follow complicated zigzags to reach deeply inside the female's reproductive tract. Damselfly penises, for instance, carry fanciful swirls and horns adorned with backward-facing bristles. This contraption allows a male damselfly to clear out 90–100% of the sperm deposited from earlier matings before passing along his own. Male penises in each damselfly species have evolved different shapes that can snake neatly through the reproductive tract of their respective females.

Most male birds don't have penises, and so they resort to some odd behaviors to safeguard their paternity. Dunnocks are rather drab sparrow-sized European songbirds. During the breeding season, female dunnocks often go around soliciting matings from different males. So a male dunnock pecks at the hind end of his nesting partner until she ejects a droplet containing rival males' sperm; only then will he copulate.

Sperm competition forces a male to simultaneously wage war on two fronts. Even after a male has mated and successfully transferred his sperm, he must still fend off future rivals. Now on the defensive, a male needs to keep the female from mating again, because these subsequent males might jeopardize his paternity. As described in the previous chapter, some *Photinus* fireflies mate from dusk until dawn. For much of this time, males are engaged in copulatory mate guarding, a strategy that prevents females from remating that night. But the stamina of male fireflies pales in comparison with that of male stick insects. Skinny they may be, yet these indefatigable guys can remain coupled to their female partners for a record-breaking seventy-nine days! Other insect males resort to chemical chastity belts to ensure they'll get to fertilize the female's eggs. When they mate, male *Heliconius* butterflies deliver an odoriferous bouquet to females, a long-lasting anti-aphrodisiac that warns other lads to keep off the lass.

Because postcopulatory sexual selection typically takes place inside the female, it's logical that females can influence males' paternity success. Drawing the analogy with more easily observed mate choice, scientists call this "cryptic" female choice. Studies in beetles, crickets, and spiders have shown that females can indeed exert control over which male's sperm gets transferred, stored, and used for fertilizing their eggs. And females can also choose to quickly remate with a new male. Although cryptic female choice hasn't yet been demonstrated in fireflies, research we've conducted in my laboratory on other beetles shows that females can bias the paternity of their offspring to favor males who are strong and healthy.

So our discovery of female promiscuity in *Photinus* fireflies had far-reaching consequences. Firefly sex was getting increasingly enigmatic. We could no longer hope to understand firefly sex simply by watching what happens in the wild, because sexual selection continued beyond mating. An amorous firefly male must not only win his lady's heart, but also ensure that he'll father her offspring. And within hidden recesses of the female's body, many things could transpire.

## Amorous Bundles

Clearly, I'd need to get intimately familiar with fireflies' private parts—unexplored territory, since no one had ever thought to look carefully inside a firefly. I soon found myself sitting at a dissecting microscope, postcopulatory sexual selection in mind. I'd spend the next few weeks poking around inside some fireflies and inspecting their genitalia. Although I didn't find any scoops or brushes for removing rival males' sperm, I did stumble upon a discovery that would forever change our view of firefly sex.

Exploring the microscopic landscapes of firefly reproductive anatomy turned out to be tremendously entertaining. This work required nothing more than open-minded curiosity, microsurgery tools, and very steady hands. I vividly recall my first glimpse into the hitherto-hidden world of firefly innards. Sunlight was streaming through tall windows in my third floor laboratory, and Alison Kraus was playing on the boom box. It was like exploring an unfamiliar yet inviting house: you climb the front steps, open the door and venture inside. Peering into rooms and down passageways, you start exploring the interior spaces. Slowly, you begin to decipher what transpires inside the house, piecing it together from small clues—furniture, pictures, the contents of various rooms. All these toys strewn across the floor? This is probably a kid's room. And the kitchen has a brick pizza oven? The owners must be serious chefs.

The interior spaces of male fireflies, I discovered, were jam-packed with stuff. And nearly all this stuff seemed to be dedicated to some sort of reproductive purpose. Over here were the sperm-producing testes, easy to spot as they're bright pink (for reasons still unknown). Yet this essential male organ was dwarfed by a big, twisted heap of reproductive glands. Quite conspicuous were two large glands that looked just like twin rotini; these spiral glands are still my favorites. Two more glands were convoluted, spaghetti-like tubes. Once I'd managed to untangle it, one gland was nearly as long as the entire firefly! Adding two smaller nubbins, I counted four pairs of male reproductive glands in total.

Following the twists and turns of their passageways, I could see that everything eventually spilled down into the male's ejaculatory duct. These male glands were clearly making something destined for export. So what was all this extra equipment manufacturing?

*Photinus* fireflies typically mate for several hours, during which they barely move. To find out what was happening beneath this calm exterior, I'd need to dissect some firefly pairs *in copula*. Fortunately, mating fireflies cooperated by remaining coupled when I placed them in the freezer. So I froze pairs at different time points, then carefully dissected them to get a time-lapse inside view. Like squeezing toothpaste from a tube, males were busily transferring some opaque goo from their bodies into the female (Figure 4.2). As expected, male sperm soon showed up inside the female's spermatheca, the compartment where female fireflies store sperm.

But the plumbing inside these firefly females was complicated. I'd been around the insides of many other female insects. And these firefly females had some re-

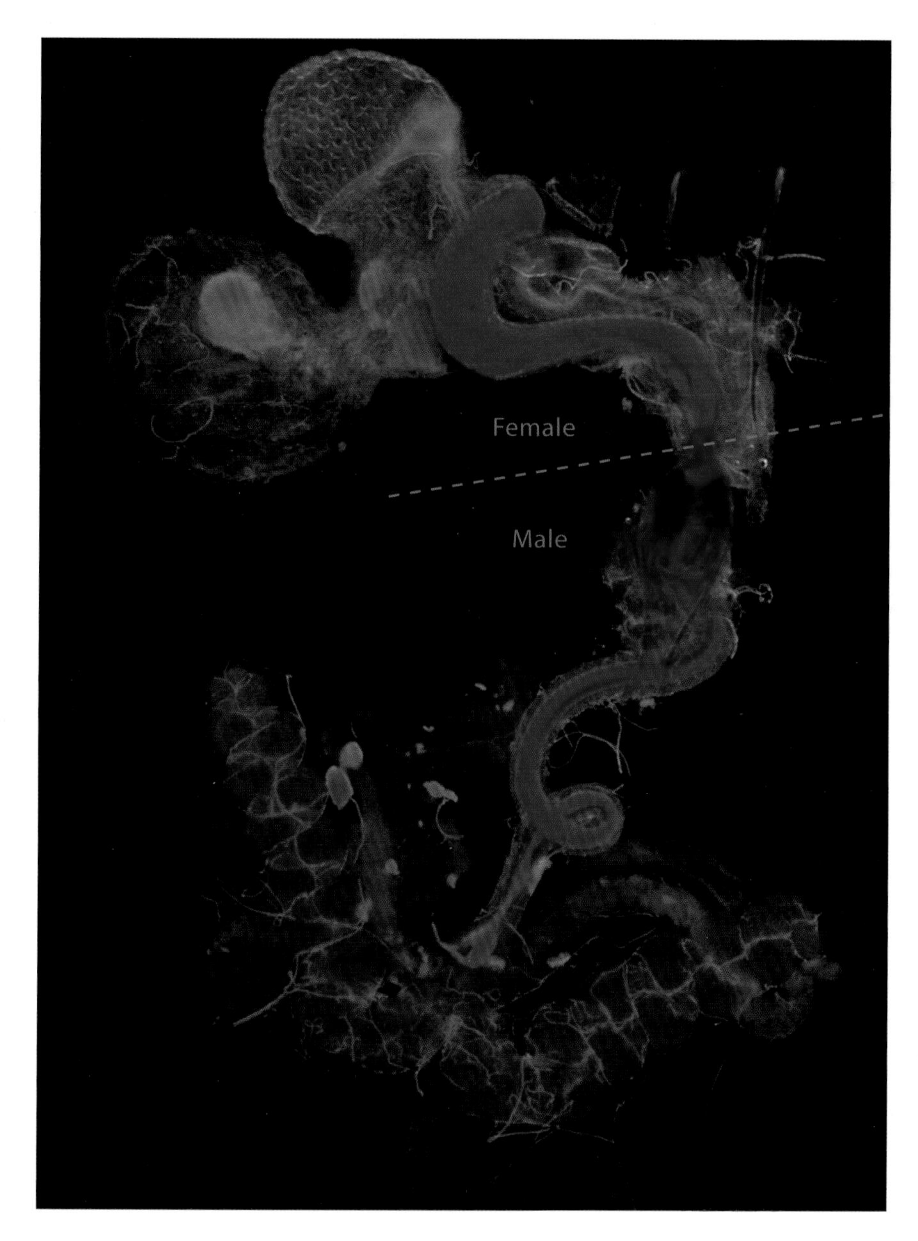


**FIGURE 4.2** While mating, a *Photinus* male gives his mate a nuptial gift, which has been stained red to make it easier to see (photo by Adam South).

productive bits that I'd never seen before, including one large, oddly elastic pouch. When mating began, this pouch looked like a deflated balloon. But an hour into copulation, a rotini-like structure appeared inside the pouch, which became hugely distended.

I'd been peering through the microscope for days, my back was sore, my eyes bleary. And then, suddenly, everything coalesced with crystalline clarity. All those

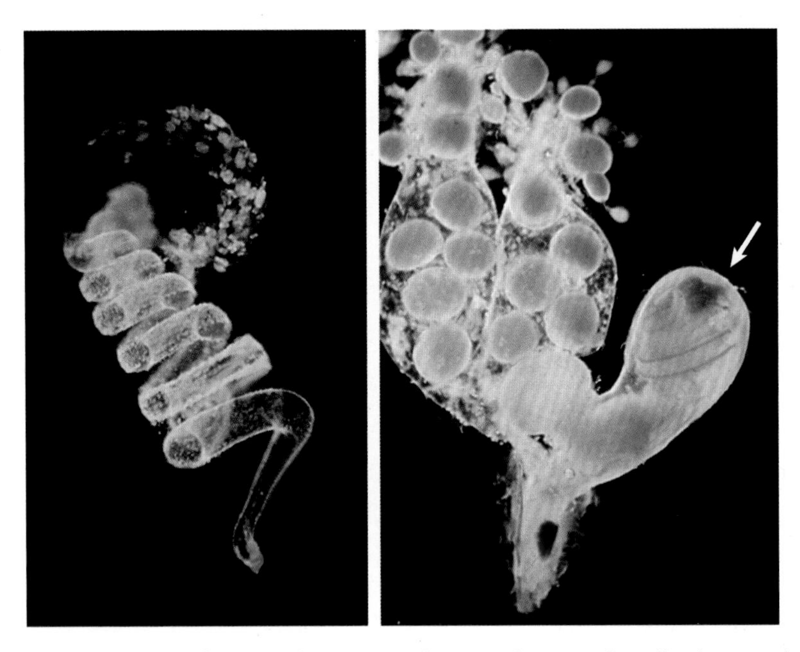

**FIGURE 4.3** Firefly bling: *Left*: A spiral spermatophore, with sperm bundles being released from the top. *Right*: Inside the female, this newly acquired male gift gets stored inside a special pouch (arrow).

male glands were busy manufacturing an amorous bundle. While they mated, male *Photinus* fireflies conveyed their sperm to the female neatly wrapped in an elegant package (Figure 4.3, *left*). The shape of this gelatinous package, known as a spermatophore, beautifully echoed the male's spiral glands. Once this bundle reached the female, the male's sperm entered the spermatheca while the rest of the spermatophore nestled into the female's pouch (Figure 4.3, *right*). Here, the male spermatophore was slowly digested over the next few days until nothing but a small, shapeless blob remained.

Firefly sex was turning out to be more seductive than I'd ever imagined! Weeks spent shining a light into dark interior landscapes had revealed something entirely new: firefly "bling." Firefly males bestow extravagant sperm packages known as *nuptial gifts* upon their mates. Although their exact biochemical ingredients are still unknown, we've learned a lot about the costs and benefits of these amorous bundles. Firefly sex is not just a simple act of gamete transfer—it's a complex economic transaction because, as we'll see shortly, these nuptial gifts turn out to be valuable commodities. But the custom of trading gifts for reproductive success is popular across the animal kingdom, so first let's see what some other creatures have on offer.

## Finding the Perfect Gift

Every single day, millions of creatures exchange gifts while they're out courting lovers and mating: humans, birds, bedbugs and butterflies, crabs, crickets and earthworms, squid, spiders and snails. In choosing the perfect gift, humans tend to pick roses and chocolate, but other animals prefer dead lizards, body parts, blood, spitballs, spermatophores, deadly chemicals, and love darts. The astonishing diversity of these so-called nuptial gifts belies their singular intent: they all serve to enhance the gift giver's own reproductive success before, during, and sometimes even after, mating.

For many species, dead prey makes a very practical gift. And for the songbird known as the northern shrike, this gift is certain to win a lady's heart. In exchange for copulations, the male shrike hunts down various prey—like voles, mice, rats, lizards, and toads—which he lovingly impales on thorns, then offers to his sweetheart.

Another male might prefer to give something he's made himself. Male scorpionflies use their gigantic salivary glands to manufacture spitballs, which females quietly imbibe while the male copulates. When his spit runs out the male scavenges a dead insect, offering that instead. Some males give rather freely of themselves. The male striped ground cricket allows a female to chew open a spur located on his hind leg; then he mates while she laps up blood oozing from his wound.

Male nursery web spiders find it expeditious to pick up something locally then add in something homemade. First they capture an insect and then spin a silken gift-wrapping around their freshly killed prey. During courtship, the male graciously offers this neat parcel to his prospective mate—how about a dinner date? When the female accepts and starts devouring his gift, the male copulates. Males can sometimes try to deceive a female by disguising some cheap gift—like flowers or prey leftovers—under a fancy silk wrapping. But once the female detects such paltry gifts, the romance quickly comes to an end.

Females appreciate these edible gifts because they're tasty and nutritious. From the male's perspective, a well-timed gift can greatly enhance his ability to win a mate. A hefty gift will also help him copulate longer and transfer more sperm, which ultimately enhances his odds of fertilizing the female's eggs.

But some gifts get delivered without the female ever having laid eyes, or lips, on them. Some males—including male crabs, shrimp, copepods, butterflies, and, of course, fireflies—invest heavily in reproductive glands that manufacture spermatophores, sperm-containing packages that get deposited inside the female. Like food gifts, these internally delivered gifts often provide nutrients that boost the female's egg production. But such gifts also include substances that quash female desire: they keep her from wanting to mate again with any competing males. This reduces sperm competition, so these gifts safeguard a male's investment by helping ensure that he'll father the female's offspring.

Beyond mere nutrients, some nuptial gifts also provide chemical bodyguard protection for the female and her eggs. The ornate moth has bright red-orange wings heavily speckled with black and white spots. Like the bioluminescence of larval fireflies, such conspicuous coloration signals to potential predators that these tiny moths are toxic. And it works: spiders, birds, and bats avoid them. Where do moths get their toxins? Ornate moth caterpillars sequester defensive alkaloids from the plants they eat, then bring along these intensely bitter chemicals when they metamorphose into adults. But sex has its own chemistry. Male ornate moths have reproductive glands that specialize in concentrating these chemicals, so their spermatophore gifts supply chemical weapons as well as nutrients. And female ornate moths rely heavily on these gifts; they'll often mate with ten or more males, accepting gifts from each one. The female keeps some of the male-donated alkaloids for her own protection—the rest goes into protecting her eggs against predators.

Lest you think all this gift giving is getting overly romantic, please be assured that it's not always so. While reproduction is generally a cooperative—and in some cases even pleasurable—venture, male and female interests sometimes conflict. So when choosing a gift, a male doesn't always have his mate's best interest in mind. Some males will hand over a manipulative gift that convinces a female to lay more eggs than is healthy for her. Some gifts hijack the female's control over whose sperm gets to fertilize her eggs. And some gifts suppress the female's desire to mate with anyone else, robbing her not only of mate options but also of gift nutrients. The upshot is that not every gift is a welcome one—we've all been given gifts we'd rather return.

Snails exchange nuptial gifts that illustrate quite nicely why it's sometimes better to give than to receive. If you're like me, then you probably haven't given much thought to snail sex. But maybe we should, because snails can get pretty kinky. Snails dispense with the usual male-female dichotomy, being simultaneous hermaphrodites. Because each snail produces both sperm and eggs, they often engage in what's known as reciprocal mating. This generally begins with a bizarre behavior known as dart shooting. When two snails come a-courting, one stabs a "love dart" deeply into its partner's body. Coated with a drug-laced mucus, this nuptial gift suppresses how much sperm the recipient will transfer the next time it mates, effectively dimming its hopes for future fertilizations. At the same time, the gift induces the recipient to store lots of the shooter's own sperm, maximizing the chances the recipient's eggs will get fertilized by the shooter. So this gift mainly benefits the giver, stacking the reproductive deck in the shooter's favor.

Personally, I find nuptial gifts endlessly fascinating. How did such a bewildering diversity of outlandish behaviors, eccentric habits, and fantastic body parts come to evolve? Why do certain creatures bestow nuptial gifts, while even their close relatives abstain from gift giving? Scientists are still trying to answer these questions, and our studies looking at the costs and benefits of firefly gifts have provided some enticing clues.

## Male Sexual Economics

Let's first take a look at gift giving from a male firefly's perspective. We've seen that fireflies—like many insects—stop eating once they've become adults. So both sexes need to fuel their reproductive obsessions using whatever capital they've accumulated during their larval feasting. Ecologists refer to such organisms as *capital breeders*. In addition to supplying energy for their nightly courtship flights, males use their stored resources to support their gift-giving generosity. How does a male firefly manage his gift export business, and what kind of benefits might he gain?

One summer we decided to ask how many times a male firefly could possibly mate if we gave him nightly opportunities. We also wanted to test the notion that spermatophores were costly. In meadows near Boston, the mating season of *Photinus ignitus* was just about to begin. We tromped out to the field with our stopwatches and headlamps and collected some early-season fireflies. Keeping males in mesh containers out in the field, we set up what we later called the wet dream experiment: each firefly male got a brand-new female delivered to his container every night. If the male mated, we collected and measured his spermatophore size. We discovered that males eagerly kept mating with these new females: the record-breaking male mated with ten different females over fourteen nights. He

**FIGURE 4.4** Firefly gifts represent a hefty investment, as illustrated by the large spermatophore this *Photinus* male has produced (photo by Wilson Acuna).

was quite impressive! And these males gallantly continued to bestow a gift every time they mated. Yet they soon ran out of steam. On average, a male's second gift was barely half the size of his first: by his fifth mating, he could only offer a tiny one-quarter-sized gift. So while a male could keep turning out spermatophores, these kept shrinking with every mating. We also learned that after they had mated a few times, males took longer to manufacture their next spermatophore.

Making gifts was clearly costly for firefly males (Figure 4.4). But nuptial gifts have persisted through evolutionary time, so they must provide some advantage. Perhaps, we thought, larger gifts might give males a leg up in postcopulatory sexual selection, allowing them to outcompete rival males and sire more offspring.

To test this idea, we'd first need to alter males' gift size, then get females to mate with two different males, and finally measure each male's paternity success. Working in my laboratory in Barnum Hall at Tufts, my PhD student Adam South eagerly undertook this complex experiment. Raised on his family's farm in Indiana, Adam was no stranger to hard work. He also knew how challenging animal husbandry could be. We'd already learned how to get big or small spermatophores from firefly males; the spermatophore produced during each male's very first mating was predictably twice as large as his second one. And we'd already worked out methods for firefly paternity testing by matching up offspring's DNA pattern with the DNA pattern of different males.

During firefly season, science is not just a day job for us, and it's not just a night job—it's a day-and-night job. We run most experiments in the Lewis Lab Flash

Room, basically a windowless closet where we have lights on timers to reverse the natural day-night cycle. Inside our Flash Room, dusk falls around ten in the morning and the sun rises at midnight. By convincing our Flash Room fireflies that day is night, we can run experiments during the daytime and still do fieldwork on wild fireflies at night. Each summer, we definitely burn the candle at both ends.

So one summer Adam became both a matchmaker and a nurturing surrogate dad to *Photinus* fireflies. He orchestrated his experiment so that each female would mate twice: once with a large-gifting male and once with a small-gifting male. After each female's second mating, Adam kept her in a moist container with some moss I'd gathered from my backyard, a special kind that female fireflies prefer for their egg laying. Every day, he carefully collected the egg-laden moss from each female, tagged it with a family ID number, and placed it in a warm incubator.

A few weeks later, Adam was thrilled to become surrogate father to more than 650 baby fireflies. These newly hatched larva came from thirty-six different families, and for each one Adam knew exactly who its mother was. By running a DNA paternity test, he could also determine which of the female's two mates was each baby's real father. (Whenever we collect fireflies from the wild for our lab experiments, we always return some eggs and newly hatched larvae back to the field site to replenish firefly populations.)

It took nearly a year before Adam finished all the paternity tests and data analysis, but in the end the evidence clearly supported our initial hunch. Firefly males that gave larger gifts were up to four times more likely to sire every single one of the female's offspring. Firefly gifts balance their high manufacturing costs by providing a reproductive benefit: they help a male to safeguard his paternity.

## Bright Lights and Bling: What's in It for Females?

Early in 2014, the Internet was abuzz with the discovery of some unusual sexual equipment: in *Neotrogla*, a Brazilian cave-dwelling insect, it's the females that have a spiny, penis-like sexual organ. These bark lice engage in leisurely matings, during which the female's organ penetrates deeply into the male's genital chamber. Then the female's organ inflates, its spines locking the pair together for up to seventy-three hours. Science reporters were pretty excited about this female penis, yet they missed the most remarkable part of this story.

Bark lice don't find much to eat in their caves. But male *Neotrogla* turn out to be prodigious gift makers, churning out large and nutritious spermatophores. The female penis acts like a vacuum cleaner, entering the male to grab his spermatophore directly from the source. So, it seems, the extraordinary sexual equipment of bark lice females has evolved simply to commandeer these valuable male gifts.

Not everybody lives in a cave, but gift giving also figures prominently in the sexual economy of female fireflies. Because these females stop eating once they become adults, making eggs is a difficult task. Each egg must contain all the nutrients necessary for development until the embryo becomes a self-sufficient, feeding larva.

So are male nuptial gifts really valuable for female fireflies? One July my graduate student Jen Rooney decided to do an experiment to test the notion that gifts contribute to females' nutrient budgets. Out in the field at night, Jen collected some early-season fireflies and brought them back to the lab. By the time she'd finished weighing and placing each firefly into its own labeled container, it was nearly two in the morning. And Jen was back in the lab bright and early the next day to divide the females into two groups: in one, each female would mate only once, while females in the second group would mate with three different males on consecutive "nights" in the Flash Room. After they'd mated, Jen indulged each female with just the right moss and egg-laying conditions. When all the eggs were finally counted, Jen discovered that male nuptial gifts help females produce more offspring; females who'd mated three times laid nearly twice as many eggs over the course of their lifetime compared to singly mated females.

Later we did a similar experiment with a different *Photinus* firefly, this time manipulating gift size rather than gift number. Females who got bigger gifts tended to live longer. So it looks like male gifts carry dual benefits for a female firefly: larger gifts bring longer life, and getting more gifts brings more offspring.

After they mated, we'd seen that female fireflies neatly tucked the male's spermatophore inside a special pouch. Here, these gifts disappeared within a few days. But where did they go? Did females eject male gifts once they'd been emptied of their sperm, as some insects do? Were the spermatophores broken down and then reused by the female for bodily maintenance? Or maybe the spermatophores were used to provision the female's eggs? We were especially interested in knowing what happened to the gift's protein, which we knew to be especially valuable because females require a lot of protein to make their eggs.

To trace the fate of the male's spermatophore, Jen used a nifty trick with tritium; this slightly radioactive but safe isotope of hydrogen would let her trace where the male proteins ended up. She started by carefully injecting some *Photinus* males with a mixture of tritium-labeled amino acids, the building blocks of proteins. Within a few days these amino acids were incorporated into the males' spermatophores. When each male later mated, he transferred a spermatophore containing tritium-labeled protein to the female. Over the next two days, Jen would trace where the male's protein traveled using a scintillation counter to measure the radioactive tritium in different parts of the female.

Immediately after the fireflies mated, all the tritium remained inside the female's spermatophore-digesting pouch along with the spermatophore. But as the spermatophore started to disintegrate, the male's protein started to show up within the female's eggs. By two days after mating, over 60% of the male-donated protein had made its way into the female's eggs. Jen's experiment demonstrated that *Photinus* females make good use of protein from the male's gift to help provision their eggs.

So gifts from male fireflies clearly represent a valuable commodity for females. And, as the previous chapter described, studies by my graduate student Chris Cratsley and others had already taught us that firefly females decide who they'll mate with based on males' flash signals. Wouldn't it be handy if females could use a male's flash signals to predict which males have the biggest gifts to offer?

A great question, but answering it would turn out to be tricky. First we had to record males' flash signals. This needed to be done in the Flash Room under carefully controlled conditions. Then we'd have to convince the same males to mate so we could collect and size up their nuptial gifts.

Chris was both brave enough and patient enough to accomplish this. Plus, we were lucky that summer to have several enthusiastic Tufts undergraduates helping out with our project. Together, Chris and these students spent many nights inside the Flash Room. They managed to coax lots of *Photinus ignitus* males into giving their single-pulsed courtship flashes, which they recorded with an extremely temperamental photometer, an instrument that uses a highly sensitive photocell to measure light. It was frustrating work. Usually when the firefly males were cooperating, the photometer wasn't, and vice versa.

By summer's end, however, the team's efforts had answered our question about whether females can judge a male's gift size by checking out his flash signals. For *Photinus ignitus* fireflies, the answer was yes, because males who gave longer-lasting flashes also gave larger gifts. This could explain how female preferences might have originated. Any females who liked longer-lasting signals would have obtained bigger gifts, which would have helped them make more eggs and leave more descendants.

But wait, not so fast! Later we repeated this experiment with *Photinus greeni*, a closely related firefly whose males court using a double-pulsed flash signal. We knew that these firefly females preferred males who flashed with shorter intervals between the two pulses. Yet in this firefly species we found that males' flash signals weren't related to the heftiness of their gifts. So maybe female fireflies do scrutinize males' flash signals trying to figure out which males are hiding the biggest nuptial gifts. If so, sometimes it will work, but sometimes it won't. As is so often the case in science, it depends.

* * *

Years worth of round-the-clock summertime explorations have helped my Tufts research team uncover many of fireflies' most intimate sexual secrets. For fireflies, mating is more than a convenient way to unite gametes—firefly males are enabling females' reproductive mission. Nuptial gifts make a vital contribution to firefly economics. Most firefly adults have stopped eating, and as females spend down their stored capital they come to rely more and more on the nutrients that male gifts provide. By season's end, as we saw in the previous chapter, the traditional script for firefly courtship gets reversed: late-season females turn competitive, while males become choosier. Finally we'd discovered the explanation: females start competing for mates because they're desperately seeking nuptial gifts.

We've learned a lot about lightningbug fireflies like *Photinus* and its relatives, where both sexes look pretty much alike. Yet let's not forget the glow-worms we encountered earlier in chapter 2—they're also members of the firefly family. Wingless and plump, glow-worm females look utterly unlike their males. Along with their distinctive sexual dimorphism, glow-worms engage in some exceptional courtship and mating rituals. They've also supplied important clues for understanding why nuptial gifts evolved in the first place. In the next chapter we'll get down with glow-worms and meet a musician-scientist who's the undisputed King of Glow.

CHAPTER 5

* * *

# Dreams of Flying

*If I had wings like Noah's dove,*
*I'd fly up the river to the man I love.*
*One of these mornings, it won't be long,*
*You'll call my name and I'll be gone.*
*Fare thee well, oh honey, fare thee well.*
- Dink's Song, Traditional -

## Into the Umwelt

*Dusk was dawning, and the forest heaved a sigh of relief. Its moist breath wrapped around me, infused with the delicious scent of earth, dead leaves, moss. Darkness eased in, carrying waves of anticipation vibrating through the air. Now it was time to switch on all eight of my lime-green lanterns. Glowing dimly at first, they soon shone brilliantly through my beautiful transparent skin.*

*Earlier, I'd shaken off my daytime drowsiness and gathered all my courage—who knew what hungry things lurked out there, waiting to grab me? I'd trudged across the lowlands for hours, then hauled myself up to this hilltop. Perched on this high vantage point, I felt I could almost touch the patchy green sky. Tonight, I vowed, I would either get myself a mate or I'd die trying. I splayed out my naked, shining body for all the world to see. I became an alluring beacon in the night, calling out "Come to me, come to me!"*

*At my eye's edge, I glimpsed a bright speck in the distance—the love of my night approaches! He zoomed over and jittered closer. Then he shone his spotlight down onto me. An electric thrill passed through me! My lanterns were burning so brightly with my desire for him, I felt I could spark a forest fire. But wait, he's circling above me. Now his light is fading away—is he leaving me? I want to shout out "Wait! Wait for me, I want to fly with you," yet I can only silently curse my earthbound existence.*

*I've dreamed of flying since I first hatched out of my egg. As larvae, all my friends shared the same dream. We excitedly looked forward to pupating, to finally inheriting our true beetle birthright: sheathed wings that would carry us high into the dark skies. But when the time finally came and I crawled out of my pupal skin, I was shocked. As we milled around that first night, I could see that only half of us—all the boy larvae—had gotten wings. And they were dark and handsome. Next to them, my sisters and female cousins all seemed so pathetic—our skin was pale and our wings nonexistent. What a letdown! Admittedly, our beautiful jewelry—those shining glow spots that covered our body—gave some consolation. But somehow it didn't seem quite fair.*

*I still remember the night the males took off, their wings lifting them up into the air as they made their bachelor flights. Some nights ago when I was young, I imagined that if I only wished hard enough, I too could snap the bonds that tether me to this earth. I'd sprout wings and climb up into the sky, watching the ground fall away below me. Maybe I could even soar above the patchy green to reach the blue layer far beyond!*

*But now I'm feeling rejected, dejected. Why did that stuck-up male ignore me? He couldn't have thought me too skinny—I'm quite attractive, plump even. Actually, my abdomen is almost bursting with eggs and it's getting difficult to move around. Really, I've just got to mate tonight and get on with laying these eggs.*

*Finally, I see more flickering lights appearing over the horizon, coming closer. Hooray, more handsome prospects! No, that one glowed too dimly, I'm glad he's passed me by. Ooooo, here's a bright fellow streaking past—I shine out "Come to me, come to me!"*

* * *

I'd done it. It was June 2013, and I was deep in the Smoky Mountains, my hair spiky with leaf litter. I'd been trying to get up close and personal with the tiny blue ghost firefly, formally known as *Phausis reticulata*. This enigmatic firefly lives mainly in the moist forests of the southern Appalachian Mountains, though it's been spotted as far west as Arkansas. Male blue ghosts float slowly just above the forest floor, glowing with a flickering eerie light. Their mystical appearance draws many firefly tourists to places like North Carolina's DuPont State Forest. Mesmerized by these floating lights, few people will ever even glimpse the creature I'd become obsessed with: the glowing, wingless female of these blue ghosts.

I'd spent the past three nights pondering these creatures, trying to get inside the mind of a female firefly who can only dream of flying. I'd wandered into a leafy glade and stretched out on my back, gazing up into the tree canopy as it faded into the darkening sky beyond. Soon, the shimmering glows of blue ghost males filled my vision. The air was filled with male lust. It was strangely titillating to have the spotlights of these tiny males playing over my supine body. At last, I'd finally managed to enter the Umwelt of a female blue ghost!

Umwelt isn't a place—it's a point of view. Early in the twentieth century, the Estonian biologist Jakob von Uexküll floated a simple but profoundly mind-blowing idea. He pointed out that different creatures, even those living side by side in the same habitat, don't uniformly share one experience of the external world. Instead, the world each animal perceives—its Umwelt—is created by its unique sensory system. This sensory filter determines what part of the world we let in, and it's been sharply honed over evolutionary time to deliver only the most relevant information. Such heightened awareness has evolved to detect only what's essential for each animal's survival and procreation: food, shelter, predators, mates.

We humans typically assume that what *we* perceive—*our* Umwelt—constitutes some objective reality. It takes practice to glimpse these other points of view, to take a giant step outside our own sensory mindset. Back in 1934, Uexküll suggested we might benefit from embarking on

> a stroll into unfamiliar worlds; worlds strange to us but known to other creatures, manifold and varied as the animals themselves. The best time to set out on such an adventure is on a sunny day. The place, a flower-strewn meadow, humming with insects, fluttering with butterflies. Here we may glimpse the worlds of the lowly dwellers of the meadow. To do so, we must first blow . . . a soap bubble around each creature to represent its own world, filled with the perceptions which it alone knows. When we ourselves then step into one of these bubbles, the familiar meadow is transformed. . . . A new world comes into being.

Although I'd set out to enter the world of a flightless female firefly, Uexküll chose to explore the Umwelt of a female tick. Sooner or later everyone encounters these unsavory parasites, which must feed on mammalian blood to complete their life cycle. Entering their Umwelt takes some effort, but if you close your eyes and block your ears, you can approach the threshold. After a female tick has mated, this "blind and deaf highway woman" climbs out along a branch at the forest's edge. Poised at its tip, she hangs motionless and waits to intercept a passing mammal. She is ignorant of everything that we normally sense, yet her world is not impoverished. Instead, it brims with incoming data from just three sensory channels. First is her acute olfactory sense, which is minutely sensitive only to butyric acid, a chemical found in mammals' sweat. So when she smells approaching prey, she abandons her perch and drops onto its back. Now her finely tuned thermal sense helps evaluate her landing spot. She thrills to temperatures of 37°C, a typical mammalian body temperature. If she's landed on a suitable host, her third, tactile sense compels her to burrow through fur until she locates a warm membrane. She sinks her mouthparts into the mammal's skin and takes a long, slow drink. At last, replete and shiny, she drops off, lays her eggs and dies.

Within its Umwelt, Uexküll also realized, time will also pass by differently for each animal. Time is perceived in a series of moments, the "briefest time units within which the world shows no change. For the duration of a moment, the world stands still." Common houseflies perceive changes in their visual field on a much finer timescale than we humans do—this is how they can so maddeningly avoid getting hit by our rolled-up magazines. Because houseflies experience more moments per second, time zooms by more quickly for them. But a tick can lurk for years before a suitable host appears. While it's waiting, the tick perceives no change in its world: no butyric acid, no 37°C, no fur. In the tick's Umwelt, each moment lasts for years—time creeps along at a glacial pace. I haven't yet figured out how time passes for fireflies. Maybe each day lasts forever, while their nights, generally frenetic with excitement, flicker by in a flash.

## Sexual Dimorphism: What's Happened to Your Wings!?

The most common fireflies in North America are lightningbugs, which court using quick, bright flashes. But glow-worm fireflies—whose flightless females use long-lasting glows to attract males—are rarely seen here. This helps explain why the blue ghost females had so vividly captured my imagination. Until I visited Thailand in 2008, I'd never seen a flightless female firefly, even though for nearly thirty years I'd been studying US fireflies. Naturally I'd read about them, but when I finally held one in my hand, I actually shrieked: "Oh no! What's happened to her wings?" The creature I took for a grotesque anomaly was actually a female *Lamprigera tenebrosus* firefly. The size and shape of my thumb, this was one giant, glowing mama of a firefly!

The most extraordinary thing about this female wasn't her huge size. No, it was the fact that she bore so little resemblance to the male who was happily mounted upon her back (Figure 5.1). More than ten times his size, her large pale body showed not the slightest vestige of wings. By contrast, sprouting from the back of the male's sleek dark body were a lovely pair of normal-looking wings and wing covers. Viewed side by side, these fireflies eloquently embodied a phenomenon known as *sexual dimorphism*, where the two sexes differ so obviously in their physical appearance that even humans can easily tell them apart. Among the world's fireflies, such sexually dimorphic glow-worms are actually pretty common: nearly a quarter of all firefly species have females who, lacking functional wings, will never be able to fly.

These oddly mismatched couples crop up everywhere in the animal kingdom. Often, it's the males whose bodies have been startlingly modified, sexual selection having furnished them with formidable weapons like antlers for defeating rivals or swank plumage for seducing females. And sometimes, sexual dimorphism shows up as a mere size difference. In such cases, it's often the female who gets the larger body size—perhaps because bigger females can churn out more eggs.

Deep-sea anglerfish provide a classic story of sexual dimorphism, and since the plot involves bioluminescence, I can't resist telling it here. Swimming slowly

**FIGURE 5.1** Large and wingless, a female giant Thai glow-worm forms an odd couple with her much smaller mate. This female has laid many pearl-like eggs, each measuring about 4 mm (*Lamprigera tenebrosa* photo by Supakorn Tangsuan).

through the ocean depths, these large, fierce-looking females attract their prey by dangling a glowing bioluminescent lure. Young anglerfish males are forty times smaller than the females, and they have huge nostrils to detect the pheromones that females release into the water to attract mates. But there are many different anglerfish in the sea. Males seem to use their outsized eyes to detect females of their own species, because different species' lures emit distinctive patterns of light. These anglerfish seem to rely on specific bioluminescent signals to find the right mate, just like fireflies.

Once the male anglerfish reaches his destination, he sinks his hook-like teeth into the female's belly. Now he'll never leave her side, because his tiny body will meld permanently with hers. His sensory and digestive systems will degenerate, and he'll draw what little nourishment he needs from the female's circulatory system. Reduced to a mere appendage, he'll dedicate the rest of his life to releasing sperm whenever the female releases her eggs.

While this anglerfish strategy of sexual parasitism might seem extreme, sexually dimorphic fireflies also divvy up their reproductive responsibilities. In glow-worms, the two sexes step out onto diverging paths as soon as the body-rebuilding process of metamorphosis begins. Males concentrate on assembling elaborate flight machinery—muscles, wings, and protective covers. Females skip all that: in some species females develop only tiny stubby wings, while other females forego wings altogether. Instead, glow-worm females channel all their energy into making eggs and building a lantern whose glow will enable them to attract a mate.

As adults, the two sexes continue to strictly divide their reproductive tasks. Emerging males take to the air, and will spend their nights hunting females. Males travel far and wide—they get to see the world. In many glow-worm species, the males don't even have lanterns, though males of other species glow while they're flying.

In contrast, the female glow-worm is destined to spend her life within a few meters of her birthplace. Some glow-worm females live in burrows made by themselves or by other animals. Laden with eggs, a glow-worm female must struggle valiantly each night to hoist her body onto a perch high enough to ensure she'll be visible. There she'll glow for hours, trying to attract a flying male. Once mated, she crawls back down to the forest floor to deposit her eggs.

Call me sentimental, but the plight of these wingless females seems awfully poignant. Surely these glow-worm's courtship rituals are miles apart from those

### Glow-Worm Confusion

An unfortunate cloud of confusion surrounds the common name "glow-worm." In different countries, people use this term to label entirely different creatures—and not one is a worm! In Europe, the word "glow-worm" refers to the glowing, flightless females of true firefly beetles. It also refers to the glowing, flightless juvenile stage (larva) of any firefly. But the famous "glow-worms" of New Zealand are not fireflies—in fact, they're not even beetles. These cave-dwelling bioluminescent creatures are actually a kind of fly known as fungus gnats. Here in the United States people likewise call these glowing fungus gnats "glow-worms." And "glow-worm" is used to refer to the luminous flightless females and juveniles belonging to the beetle family Phengodidae. To distinguish these from true fireflies, which are considerably smaller, phengodids are sometimes called "giant glow-worm beetles." Needless to say, this profusion of glow-worm usage can cause a lot of confusion!

of their more egalitarian cousins, the lightningbug fireflies that we met in the previous chapter. For one thing, glow-worm females can't opt to move out of the neighborhood if males or good egg-laying spots are scarce where they live. So it's easier to accidentally snuff out a glow-worm population (some things likely to do this are described in chapter 8).

After I'd seen that giant *Lamprigera* female lying next to her tiny mate in Thailand, I began to wonder how such startling physical differences might alter glow-worms' sexual habits. Up until this point, my students and I had concentrated our studies on understanding the role of nuptial gifts in North American fireflies. We had discovered that lightningbug males provide gifts during mating, and that such gifts help their females produce more eggs. We also knew that females of all our typical lightningbug fireflies have normal wings; they're perfectly capable of flying when they want to. Because adult fireflies don't eat, all their activity must be fueled by whatever reserves they've accumulated during their months of larval feasting. Lightningbug females, then, must split their investment between flight and reproduction: if you want to fly, then you'll have fewer resources left for reproduction.

But what about glow-worm females? They've given up their wings to become waddling bags of eggs—everything they've got, they've put toward reproduction. If they've already maximized their reproductive investment, would male gifts still be so important to them? I began wondering whether females' flight abilities might influence males' gift giving. And it seemed like these glow-worm females could help us find out.

Working together with international collaborators, we gathered data on a few dozen firefly species, including many glow-worms from around the world. For each species, we noted whether females were flightless or could fly, and whether males provided spermatophore gifts during mating or not. We also constructed a phylogenetic tree for all these fireflies, using differences in their DNA sequences to trace their evolutionary history.

When we finally mapped females' flight ability and male gift giving onto this tree, we found we'd uncovered a surprisingly tight evolutionary link. Nuptial gifts are ubiquitous among those fireflies with flying females, but in most glow-worm species such spermatophore gifts are absent. This was exactly what we'd predicted, based on the difference in females' reproductive investment: the firefly gift-giving tradition turns out to be limited only to those species whose females can fly. When females channel everything they've got into reproduction—as glow-worms do—the sexual balance shifts, and males lose their ability to produce spermatophore gifts. In the face of such total maternal devotion, firefly males have apparently decided that such gifts are no longer worth giving.

These results may help explain how nuptial gifts evolved in other animals as well, and we published our study in 2011 in the scientific journal *Evolution*. Yet my infatuation with glow-worms was just beginning. I'd soon find myself bewitched by the love lives of blue ghosts, a native North American glow-worm.

## The King of Glow

In many European countries, the night of the summer solstice happens to coincide with peak adult glow-worm activity. This holiday, known as Saint John's Eve, is celebrated with bonfires, dancing, and other nocturnal merriment. Traditional stories and folklore surrounding these festivities tell of fairy gatherings and of plants acquiring mystical powers, a magical aura immortalized in Shakespeare's *A Midsummer Night's Dream*.

During their nocturnal escapades, more than a few solstice celebrants have probably stumbled, wide-eyed, upon a dark, forested place with hundreds of tiny lights sprinkled over the ground. Not fairies, these magical glows would be females of the common European glow-worm, *Lampyris noctiluca*. In northern European latitudes, these females begin shining at dusk—around the summer solstice, dusk falls just a few hours before midnight. A female might glow for two hours or more, rain or moonshine, and her glow is visible up to 50 meters away. Up on her perch she does a slow, seductive dance, swinging her lantern from side to side. This rhythmic dance will help her stand out to the glow-worm males who are searching for her while they fly, unlit, through the dark forest.

I first met the King of Glow in 2008, the very same year I encountered my first flightless female firefly. Raphaël De Cock is a shy, handsome Belgian who leads a double life. While he now earns his living as a well-established folk musician, he is also recognized as a world expert on European glow-worms. Somehow Raphaël deftly juggles this dual-career life, though he's met some challenges along the way.

Both of Raphaël's passions surfaced early on. Born in Antwerp in 1974, he grew up spending weekends at his grandparents' house in the countryside. There he fell irreversibly in love with all things that glow. With a faraway look, Raphaël recalls, "My grandfather had a book filled with photos of different bioluminescent creatures on land and in the sea. Whenever I opened it, I stepped into a mysterious, fairy-tale world." During his childhood Raphaël amassed a collection of fluorescent minerals, admiring them under his black light as they transformed into a glowing rainbow of colors. As a boy, he also became entranced by phosphorescent glow-in-the dark toys that, once charged, slowly reemit the light they've absorbed. (Just as the moon shines only with reflected sunlight, phosphorescent and fluorescent objects don't shine with their own light—they must first be illuminated.) Even now, as an adult, he readily admits to a "strange fetish for collecting glow-in-the dark toys." He fondly recalls his glow-in-the-dark Playmobil ghost from his childhood—"This one was definitely my all-time favorite! In fact, I still have it." Like a bowerbird, he surrounded himself with glow toys in every available color and shape—stones, lizards, putty, and stars.

His grandfather's book also inspired Raphaël's earliest firefly explorations in the Belgian countryside. At the age of nine, he discovered his first glow-worm larva hidden beneath a rock. Elated, he placed it carefully on a leaf and ran quickly home to show his grandparents. When he arrived the larva had fallen off its leaf somewhere—Raphaël was devastated. But soon bioluminescence would leap to life right off the book's pages, ushering him into its hidden world. He tells of one night out exploring with his grandmother: "We were walking slowly in total darkness, letting our eyes get dark-adapted, when suddenly I saw so many glowing dots moving across the forest floor," he says, explaining excitedly, "They were all glow-worm larvae!" Raphaël recognized these as *Lampyris noctiluca* from the book's photographs. He brought some larvae home and set them up in his room. He gathered snails to feed his bioluminescent pets and then watched in amazement as they turned into glowing pupae. From then on, a box of glow-worms accompanied him on every family vacation. But he hadn't yet become the King of Glow.

When he entered the University of Antwerp, his glow-worm curiosity drove Raphaël into scientific research, and he stayed on to earn his PhD. His boyhood observations had taught him that firefly larvae emit light when he picked them up, or even stomped on the ground near them, and they also glow spontaneously while they crawl along. But here's the question he really wanted to answer: *Why* did these firefly larvae glow? What, exactly, is the point of being so conspicuous? After all, it seems to be inviting trouble—"Here I am, eat me!" Maybe glows helped the larvae find their way? But this seemed an unlikely explanation, because firefly larvae have lousy eyesight. Attracting prey? This also seemed unlikely, since glow-worm larvae get prey by actively chasing down snails.

Raphaël had read earlier studies reporting that fireflies taste awful—toads, birds, lizards, and many other predators wouldn't eat them. So he wondered: might the lights of glow-worm larvae act like a signal to ward off nocturnal predators? Raphaël's experiments during his PhD research provided a key piece of evidence supporting the idea that firefly bioluminescence originated as a warning signal to fend off potential predators. These now-classic experiments demonstrated that nighttime predators, like toads, can learn to associate a glow signal with noxious prey; following an initial unpleasant encounter, they stop attacking similar-looking prey.

At last, the King of Glow had ascended the throne. Raphaël had earned his doctorate, enriched our knowledge about firefly warning signals, and written many scientific papers. In one paper, he showed that females of the lesser European glow-worm (*Phosphaenus hemipterus*) attract their mates by emitting alluring perfumes, the chemical signals known as pheromones. A decade later, this discovery would convince us that we should mount a joint field expedition to the Smoky Mountains in Tennessee, to see whether blue ghost females might use similar perfumes to bring in their males.

By 2005 it looked like Raphaël was headed for an illustrious academic career. But the Belgian job market for scientists was bleak, and Raphaël soon despaired

### The Glow-Worm Song

The glow-worm's undeniable romantic charm was captured in a song "Das Glühwürmchen," which later became a popular song in the United States. Originally written for Paul Lincke's 1902 operetta *Lysistrata*, with German lyrics by Heinz Bolten-Backers, the English translation below is by Lilla Cayley Robinson. This song was performed in the 1907 Broadway musical, *The Girl Behind the Counter*. Later rewritten to retain only the chorus, "Glow Little Glow-Worm" was recorded by the Mills Brothers in 1952. My mother used to enjoy singing this to us at bedtime. Highly entertaining (though scientifically inaccurate), this song remained a hit throughout the 1950s.

When the night falls silently,
The night falls silently on forests dreaming,
Lovers wander forth to see,
They wander forth to see the bright stars gleaming.
And lest they should lose their way,
Lest they should lose their way, the glow-worms nightly
Light their tiny lanterns gay,

Their tiny lanterns gay and twinkle brightly.
Here and there and everywhere,
from mossy dell and hollow,
Floating, gliding through the air,
they call on us to follow.

Shine, little glow-worm, glimmer, glimmer
Shine, little glow-worm, glimmer, glimmer!
Lead us lest too far we wander.
Love's sweet voice is calling yonder!

Shine, little glow-worm, glimmer, glimmer
Shine, little glow-worm, glimmer, glimmer
Light the path below, above,
And lead us on to love!

of ever finding a suitable university position. So to earn his living, he nimbly switched gears to take advantage of another talent—making music.

As a young boy, Raphaël had played the recorder and tin whistle. Then, shortly after his sixteenth birthday, an early musical aptitude blazed into sudden obsession. Turning on the radio one day, he heard someone playing the uilleann pipes, the national bagpipe of Ireland. The voice of this instrument—something like a cross between a fiddle and an oboe—touched Raphaël's soul. Although uilleann pipes are famously difficult to play, Raphaël would eventually find his voice in this instrument, sweeter, quieter, more fluid than other bagpipes. Over the years, he also became adept at Mongolian throat singing and added to his repertoire a dazzling panoply of ethnic instruments.

Now Raphaël travels the world—Canada, Siberia, Bolivia, Sardinia, Ireland, Scandinavia—as a professional musician, teacher, and performer of world music. Although he's chosen to spend his life outside the confines of academia's ivory towers, Raphaël continues his scientific studies on fireflies everywhere he travels. "I should have lived in olden times," he says wistfully, "when it was normal for people to pursue many different interests simultaneously: science, music, art. In this modern world, our lives have gotten so specialized—it's a pity." Perhaps not surprisingly, Raphaël hears music whenever he sees fireflies flashing—every firefly species makes a different sound. For the King of Glow, these sparks aren't silent at all.

## Ghostly Glows and Phantom Fumes

In 2013, I was still suffering from glow-worm addiction—the bleak lifestyles of those flightless females kept haunting me. In my travels around the world, I'd seen many of these creatures, but I had yet to meet an American glow-worm crawling around in her native habitat.

Now I was in Knoxville, Tennessee, chatting with naturalist Lynn Faust in the frigidly air-conditioned airport as we waited for the last member of our research team to arrive. Raphaël De Cock stepped lightly off the plane, his insect-collecting net slung over one shoulder, musical instruments slung over the other. Together we were heading off on a field expedition into the Great Smoky Mountains, where we'd be focusing our collective scientific attention on the mysterious blue ghost firefly, *Phausis reticulata*. We'd chosen to study blue ghosts because these glow-worm fireflies were relatively local, and also abundant. Soon, though, we'd find ourselves falling under their spell.

Before we arrived in Knoxville to begin our intensive study, we'd scoured the scientific literature for everything ever written about blue ghost fireflies. They were mysterious all right—we didn't find much. *Phausis reticulata* had been formally named and described (from dead museum specimens) back in 1825, but surprisingly little was known about their courtship habits or mating rituals. Blue ghost males have tiny, dark bodies—just about the size of short-grained rice—and a light-producing organ that occupies two segments at the tip of their abdomen. During their mating season, they fly through the forest at ankle height for about two hours each night, searching for females. While they're flying, males emit a constant glow that lasts about a minute and appears faintly bluish. Many people describe their light as ghostly, hence their common name. One observer, seeing blue ghosts for the first time, thought, "Fairies . . . and they are carrying tiny blue lanterns."

Meanwhile, the wingless blue ghost females had only rarely been described in the scientific literature. Roughly matching the size of their males, their pale bodies blend perfectly into the leaf litter. At dusk, they begin to glow from the forest floor, lighting up from several luminous spots that shine right through their nearly transparent skin.

Our expedition to study blue ghosts in the wild would only last a few weeks, yet we hoped to answer several questions. Do blue ghost females attract males only by their glow, or do they have other courtship tricks up their sleeve? Do females mate just once? And are their lights really blue?

We'd gathered the perfect scientific trio for this study. Raphaël had studied European glow-worms extensively, and had discovered that one species' wingless females attract males with chemical signals. Lynn Faust knew everything there was to know about Tennessee fireflies. Riding through the mountains one night, she'd come upon a low cloud of blue ghosts so dense that her horse, Echo, was bewildered and kept trying to step up onto the luminous surface. And she'd been watching blue ghosts around her family's farm near Knoxville for more than fifteen years. I rounded out the group with my expertise about firefly sexual selection, mating behavior, and nuptial gifts. We also knew we'd get along—a good thing, since this expedition would require us to work intensively day and night, live together in close quarters, and get very little sleep.

From the airport we head straight to our study site with our headlamps, our field notebooks, and our cameras, arriving around ten at night. A breathtaking spectacle greets us when we enter the forest (Figure 5.2). Everywhere, blue ghost males are floating slowly above the forest floor, glowing. Together, they create pools of living light that gather, overflow, then cascade silently down the hillside. Wrenching our attention from the males' meandering paths, we start searching for the flightless, yet elusive, females. We crawl through the leaf litter

**FIGURE 5.2** Blue ghost males weave glowing paths over the forest floor as they search for their wingless females (*Phausis reticulata* photo by Spencer Black).

uncovering these hidden jewels, their transparent bodies spangled with tiny glowing spots.

When I slide into sleep that first night, ghostly lights still float before my eyes. And by morning, they've cast a spell that has transformed our research project. What a privilege it would be to get to know these enigmatic creatures! If only they'll grant us access, we might discover something new, utterly wondrous about their arcane courtship habits.

Early the next day, Raphaël, Lynn, and I get organized for the field experiments we'll be conducting over the next few nights. We aim to test whether female blue ghosts might be attracting males not just by their jewel-like glows but also by emitting pheromones. First, we set up temporary lodgings for the blue ghost females we'd collected the previous night. These consist of cardboard cups—we just happen to have some, originally intended for serving ice cream sundaes—that we've lined with moist paper towel and leaf litter from the blue ghosts' own forest. Gently lifting the delicate females with a paintbrush, we carefully settle each one into her own apartment.

We've designed three different lids to control exactly which female signals will escape from these containers. Will the female glow alone be sufficient to lure in a male? If a female's glow is hidden but her smell can escape, will males still come calling? To answer these questions we set up a simple experiment with our ice cream cups. Some containers get covered with simple mesh lids: both the female's glow and her smell can escape, but males can't reach her. Others we cover with airtight, clear plastic lids: the female's glow will be visible to males, but smells can't escape. On the remaining containers, we install a light-blocking baffle above the mesh: female glows won't be visible, but smells can still diffuse out.

Over the next several nights, we head into the field before dark. Blue ghost males start flying about forty minutes after sunset, but we arrive early to set things up. We set out our females, now comfortably ensconced in their ice cream containers, then retreat to the sidelines to wait quietly for complete darkness. Within a few minutes, our tiny blue ghost females have crawled up to the top of their leaf litter, and they've started to glow. They're ready! And at last, here come the males gliding silently down the hillside.

We hardly talk for the next two hours. Each of us is responsible for closely watching three different females. We stop every ten minutes, recording on our datasheets exactly how many males have passed through each female's airspace, and how many of these males have actually landed. It's midnight by the time the males have all stopped flying. We've been caught up in a frenzied blur of blue ghost courtship, and now our datasheets are full, notes scrawled front and back.

We doggedly repeat this experiment night after night, adding new females and releasing the old ones. As we crunch numbers and graph our accumulating data, a pattern slowly emerges. Numerically, the females in each experimental treatment all seem equally attractive—roughly the same percentage of passing males (between 10 and 40%) has landed to check them out. But we've been observing this blue-ghost mating scene very closely, and we've detected some tell-tale signs that females are releasing some attractive perfumes. We've seen far-off males flying upwind to reach a female, even when our cardboard hides her glow. Some males approach indirectly, tacking toward the female like sailboats headed upwind, while others seem drawn like magnets.

It was Female #3 who really convinced me there might be more to blue ghosts than simply their glow. During the first hour that I watched her, things were pretty quiet. Several blue ghost males flew past, but only one was interested enough to circle and land on her container. He crawled around, but soon departed when he couldn't get past the mesh. By ten thirty only a few males were still flying—the action was definitely slowing down. Suddenly, four males flew in from nowhere and landed right on the container! It seemed like Female #3 had just then released some secret scent. So we think these females may have a backup plan, which we called the wallflower stratagem: if their glow display doesn't bring in a mate, they'll resort to releasing pheromones in hopes of snagging a late-flying male.

For now, we need to wrap up our experiments—they'll be just the first step in exploring this fascinating world of glows and perfumes. We're hopeful that our tantalizing results will inspire others to look at glow-worm courtship with new eyes—and noses. Perhaps someday you'll even be able to walk into a store and buy blue ghost perfume!

During this pheromone-testing experiment, we've been working on the dark side of the mountain, and collecting the data on blue ghost behaviors has demanded our full attention. But as we walk back to the car after the final night of this experiment, the wider world rushes in. The moon clears the hill behind us, spilling moonglow and tree shadows across our path. The Great Smoky Mountains loom in the distance, slumbering peacefully. Now I can see that they're covered in blankets woven from the dancing, flickering lights of blue ghosts.

During daytime, too, we have lots to accomplish, so we're glad for the generous day lengths near the summer solstice. As we examine the nightly portraits we've taken of each blue ghost female, we're surprised to find different numbers of glow spots. Some females shine from only three spots, while others glow from as many as nine. Maybe larger females have more spots? One day I spend several hours carefully measuring their tiny bodies and taking photographs with a microscope connected to my laptop. Some females are triple the size of others (Figure 5.3). And it turns out these larger females generally *do* have more glow spots. This sets us wondering—are blue ghost males drawn to females who shine with more spots?

Field biologists are proud of their ability to cobble together whatever equipment they need from materials that happen to be at hand. Raphaël shows us some tiny luminescent tubes, called Betalights, that he's brought along in his traveling collection of glow-in-the-dark objects. And he's figured out how to imitate the different glow patterns we've seen on real blue ghost females. Slipping the glow-

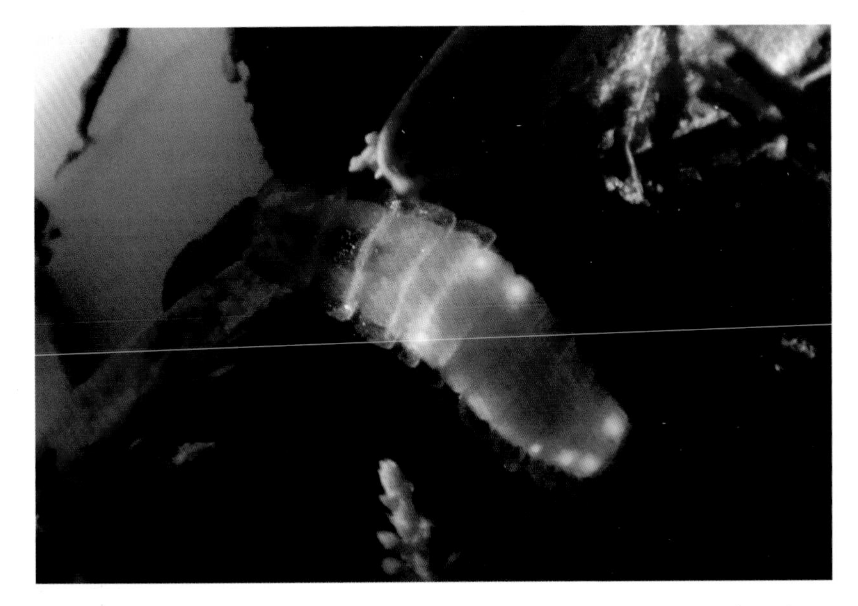

**FIGURE 5.3** With glow spots shining out through their transparent skin, the delicate blue ghost females look like tiny gems nestled into the leaf litter (*Phausis reticulata* photo by Lynn Faust).

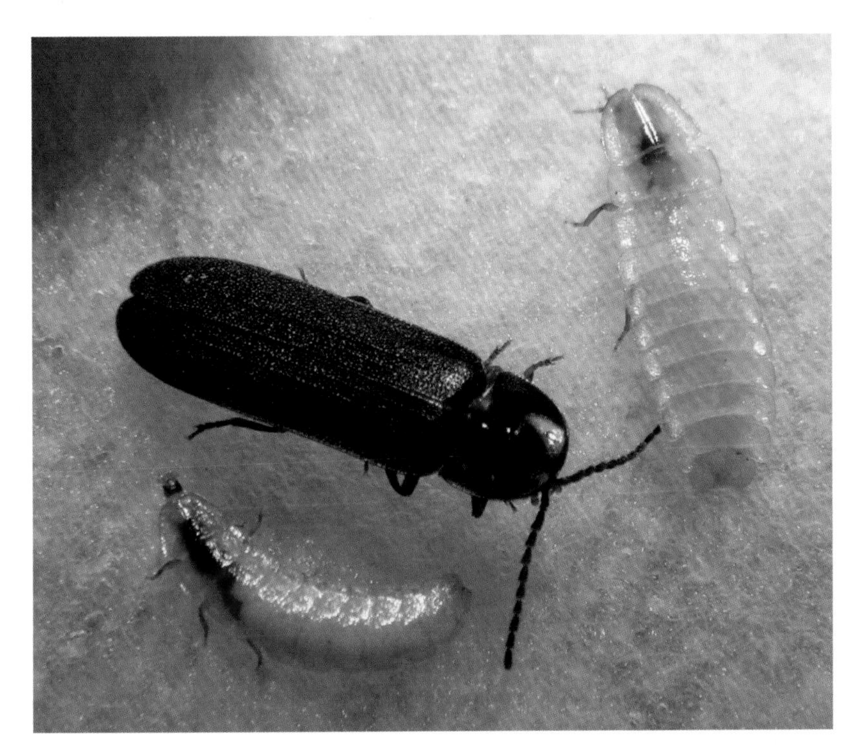

**FIGURE 5.4** A male blue ghost firefly (center, ~8 mm long) entertains two pale, wingless females (*Phausis reticulata* photo by Raphaël De Cock).

ing Betalight inside a black plastic drinking straw, Raphaël pierces the plastic with a needle, creating tiny holes where the light can shine out. He spends one entire day bent over the straws, finally producing some 4-spot and 8-spot glow lures; in a darkened room, these very convincingly mimic actual females. The next afternoon we go dumpster diving and score two dozen 2-liter plastic soda bottles. With scissors, paint, and thread, we fashion these into funnel traps, and suspend a glow lure inside each one. We're hoping that males will be attracted to the lure, enter the funnel, and get trapped inside, and that they won't be harmed. When we test these contraptions out in the blue ghost forest, we're happy to find they work quite well.

Which lures will attract more blue ghost males? For several nights, we set out Raphael's lures at dusk and collect them at midnight when the males' flight period is over. We open the funnels, count the unharmed males inside, and release them. When we tally up the totals, we find twice as many blue ghost males in the traps with the 8-spot lures. They probably don't count, but it seems that blue ghost males really do fall for females with more glow spots. We can now appreciate how this makes good biological sense: males who choose more glow spots will end up mating with the largest females, and these are the females who will have more eggs.

We each also take a few nights to indulge our personal fascinations with the blue ghost lifestyle. The King of Glow plans to find out exactly what color light these "blue" ghosts emit. To do this, Raphaël has brought along a portable spectrophotometer, an instrument that records and very precisely measures the wavelength of any light. Surprisingly, his measurements reveal that both sexes shine light that's actually lime green (its peak wavelength is 554 nanometers). This nicely matches what's been found for other glow-worms. Yet it doesn't explain why, when viewed from above, the light from blue ghost males really does seem bluish. Perhaps this color-changing illusion is from the males' glow reflecting off the dark green leaves below.

Another surprise emerges when we discover that blue ghost females are stay-at-home moms. Lynn has been keeping track of our lab-mated females, watching as each one lays her sticky clump of a few dozen eggs on the leaf litter we've provided. Insects aren't known for their maternal devotion—most female insects lay their eggs and then take off. So Lynn is astonished to see these blue ghost females slowly curl their pale bodies around their eggs, embracing the clump with their legs. Lynn gently disturbs them with a paintbrush—they glow brightly, then crawl away. Yet within minutes the moms move back into position, clutching their eggs protectively. Day and night, our blue ghost females continue their egg guarding until, about a week after egg laying, they die. Now left on their own, the eggs will take another month before they hatch out into tiny, crawling larvae. They'll never meet her, but their mother's maternal devotion provides her offspring with a head start on survival. Female of another glow-worm, the giant *Lamprigera* fireflies in Thailand, also exhibit maternal care. *Lamprigera* females curl their bodies around their eggs, cleaning them carefully every day for three months until the larvae hatch out. Along with their glow, these stay-at-home moms might possess yet-to-be-discovered chemical weapons that deter egg-eating predators and pathogens. Our admiration for these blue ghosts—and for their mysteries—is growing by leaps and bounds.

Meanwhile, I get to explore the Umwelt of a female blue ghost, described at this chapter's beginning. For a few nights, I walk into the forest, settle down in the leaf litter, and watch males fly mere inches above my nose. Our pheromone experiments added a new sensory dimension to this courtship scenario. Inside the blue ghost's Umwelt, I now imagine I can see a female's attractive scent wafting through the forest—it's fluorescent perfume made visible. Around me, other females are releasing their irresistible perfume in clouds that rise and swirl, getting dimmer and more diffuse as they mingle with the fern-flavored air.

The zoologist and Nobel laureate Karl von Frisch likened the honeybee, his favorite organism, to a magic well: the more water you draw from it, the more remains. Our field expedition, originally inspired by the sheer mystery of this firefly, has revealed some of these glow-worm's most closely held secrets. What we've learned has certainly transformed our view of these blue ghosts, yet many mysteries remain. But now I'm ready to head back home, feeling grateful for the chance to study these enchanting creatures, and to drink in their wonder.

* * *

We've heard quite a bit about *why* fireflies flash, and now it's finally time to address another big question: *how* do fireflies flash? In the next chapter, we'll open the hood and explore the nuts and bolts of how fireflies actually make their light. We'll also take a look at how this bright talent might have originated. Along the way, we'll discover that fireflies are not merely beautiful. Their light-producing chemicals have helped save human lives by enabling advances in public health, medicine, and scientific research.

CHAPTER 6

* * *

# The Making of a Flasher

*Energy is eternal delight.*

- William Blake -

## A Chemistry Set for Light

I was playing badminton without a net when I really met my life partner, Thomas. One hazy summer evening in New Hampshire, the shuttlecock whizzed back and forth between us. Suddenly, we were surrounded by sparks rising from tall grass all around us. Having been raised in New England I took this spectacle completely for granted, yet Thomas froze in amazement. He had grown up in Oregon, a place sadly devoid of fireflies; he'd only just moved to Boston to attend college. Now, as I watched his eyes widening with awe, my own eyes were remade for wonder. I felt like I was seeing Thomas, and the fireflies, for the very first time. Decades later, swapping spectacle for science, we'd find ourselves working side by side to decipher one of the mysteries of firefly flashing.

These silent sparks seem so magical, yet their lights arise from a carefully orchestrated chemical dance that takes place inside the firefly lantern. The luminous beetles—fireflies among them—head up the roster of terrestrial creatures that have learned the trick of transforming chemical energy into light (others include some fungi, earthworms, millipedes, and fungus gnats). On land, bioluminescence has evolved independently at least thirty separate times, and many more light-producing animals inhabit the sea. All these luminous creatures have come up with many different chemical systems to make their lights. Confusingly, this biochemical diversity gets hidden because scientists use the generic name *luciferase* (from the Latin *lucifer* "light bearer" + *-ase*, the suffix attached to all enzymes) to refer to any enzyme that catalyzes light production. Yet though they each catalyze a similar chemical reaction, the luciferases used by different creatures can vary substantially in their structure.

Each luciferase consists of a large protein with a particular three-dimensional shape that allows it to coax its dance partner, always a much smaller molecule, into emitting light. These smaller molecules are collectively known as *luciferins*, and they're what actually produce the light that we see. All luciferins hail from the same general class of chemical compounds; they each possess several linked rings that mix carbon, nitrogen, and/or sulfur in various combinations. Luciferin's special talent lies in its ability to trap chemical energy in the bonds between these rings. Gently manipulated by its luciferase partner, luciferin harnesses this energy to give off light.

Many secrets of how animals produce light were uncovered by studying the American Big Dipper firefly, *Photinus pyralis*. In these fireflies, the light reaction is catalyzed by a luciferase enzyme consisting of exactly 550 amino acids—these are the building blocks of protein—strung together in a chain. Along with luciferin, firefly luciferase requires just a few other players to make light. One is adenosine triphosphate, commonly abbreviated as ATP. ATP is a VIM (very important molecule)—it's how all living things transfer chemical energy around inside their bodies. Oxygen molecules—like the ones we breathe—round out the cast. All these other molecules are tiny relative to luciferase, yet each one is essential.

The players are now assembled; so let the dance begin. The light-producing action takes place within one particular hollow on the luciferase molecule, a location known as the enzyme's active site (Figure 6.1). Here, all the players can nestle together in just the right position to participate in this delicate molecular dance.

Fireflies make their light using a multistep process. In the first step, luciferase brokers an interaction where ATP can transfer some of its energy to luciferin. Because this form of luciferin is fairly stable, fireflies keep this intermediate in stock. In the second step, oxygen joins the party to convert luciferin into a highly excited chemical state. This excited luciferin is evanescent—it stays excited for only a few hundred millionths of a second. As it relaxes back down from its high-energy state, luciferin releases photons of visible light, like miniature lightning. In the final step, a luciferin-regenerating enzyme prepares luciferin to once again be

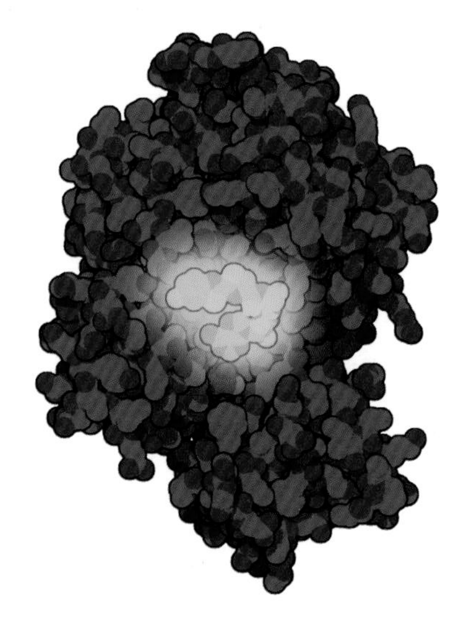

**FIGURE 6.1** Enfolded within the active site of the luciferase enzyme, a molecule of luciferin gets manipulated into transforming its chemical energy into light (image by David Goodsell).

## Tracing the Evolution of Beetles' Bright Lights

About 2,500 beetle species are capable of making their own light, and they all hail from only four (out of 176) beetle families. The vast majority of light-producing beetles belong to the true fireflies (family Lampyridae)—as we've seen, all fireflies are luminous during their larval stage. But the luminous list also includes about 250 giant glow-worm beetles (family Phengodidae), thirty species from a family known as Rhagophthalimidae, and about two hundred species of click beetles (family Elateridae).

Current phylogenetic evidence suggests that bioluminescence evolved once in some ancient progenitor that eventually gave rise to the first three beetle families, including fireflies. Light-production probably evolved separately in the click beetles, however, and only a handful (< 2%) of these are bioluminescent.

For about forty years now, scientists have been isolating and reading off the DNA sequence for the luciferase-encoding gene (called *luc*) that's found in many luminous creatures. We now know the DNA sequence of the *luc* gene from many different light-producing beetles, including about thirty firefly species. Using this information, scientists have deduced the exact amino acid sequence for the luciferase enzyme. This sequence of amino acids turns out to be more than 46% identical across all luminous beetles.

Luminous beetles have evolved to shine out a rainbow of colors—red, orange, yellow, and green. These color changes appear to be caused by small differences in the luciferase enzyme. If even a single amino acid near the enzyme's active site changes, it causes its dance partner luciferin to twist into a slightly different conformation. This produces sparks with different wavelengths, dazzling our eyes with different colors.

swept off its feet by luciferase, ready to take another light-producing turn around the dance floor.

Fireflies make their light with higher efficiency than any other bioluminescent creature: their quantum yield is around 40%. This means that 4 photons of light get emitted for every 10 luciferin molecules chemically transformed. Compared to the typical incandescent light bulb, which shines with efficiency only around 10%, this is quite impressive.

What about the dark fireflies—that is, those species whose adults can't light up? Adults of the day-active Asian firefly, *Lucidina biplagiata*, still contain both luciferin and luciferase. But now these players are present in just tiny amounts, only about 0.1% of what's found in their light-producing cousins. Since these chemicals are no longer necessary, such day-active fireflies presumably conserve energy by dialing back their production.

## Firefly Lights Evolving

How did fireflies acquire their ability to produce these silent sparks? That is, where did luciferase, the enzymatic star of this bioluminescent light show, come from?

It seems likely that an early firefly progenitor creatively improvised this key light-producing enzyme, using as starting material a fat-metabolizing enzyme. There's a striking resemblance between firefly luciferases and another enzyme family known as fatty acid synthases. Multicellular organisms use the latter to build fatty acids, and these fatty acid synthases thus play a crucial role in basic metabolism. Because these enzymes are so important, they're found in many different versions inside animal cells. Just as luciferase uses ATP to chemically transform its luciferin substrate and make light, these metabolic enzymes use ATP to chemically transform a different substrate. But it turns out that modern-day firefly luciferase, when given the right substrate, can behave just like a fatty acid synthase. This dual talent suggests that the starting point for luciferase was probably an enzyme that helped to metabolize lipids.

More experimental evidence supporting this notion comes from studies on the mealworm beetle *Tenebrio molitor*. (We'll encounter the larvae of these beetles in the next chapter, though in a somewhat more gustatory context.) *Tenebrio* is only distantly related to fireflies, although both are beetles. But when researchers took luciferin out of fireflies and injected it into a live mealworm, this normally lightless beetle gave off a faint, reddish glow. This demonstrates that even *Tenebrio* beetles have some enzymes capable of producing light, as long as they're given the appropriate light-emitting substrate, like luciferin.

Scientists think that firefly luciferase, like many other novel enzymes, originated through an evolutionary process called gene duplication. Here's how it might have happened. In the distant past, a sequence of DNA coding for some fatty acid synthase accidentally got duplicated during the process of DNA replication. Because the original gene was still competently performing its job, the extra copy could lounge around and freely accumulate mutations. Most of these mutations led nowhere, but some produced functional enzymes with distinctive new properties. One mutant enzyme even happened to produce light—at first, this was a mere by-product of some new reaction. But the light produced by this proto-luciferase conferred some benefit, so natural selection favored the spread of this particular mutant form of the duplicated gene. Over eons, selection for more efficient light production concentrated this luciferase into a specialized tissue, which eventually became what we now call the firefly's lantern.

Gene duplication creates redundancy, and redundancy provides fuel for evolutionary innovation. Because the duplicate becomes a free agent, over time it can diverge and might eventually specialize in performing some entirely new function. While it's tempting to think of evolution as goal-oriented, in reality there's no intent, no predetermined trajectory. Sometimes the result of evolution's creative improvisation is a flop, and sometimes the result is useful. This process of gene duplication likely gave birth not only to luciferase but also to many other novel metabolic enzymes during the long history of life on Earth. For instance, the enzymes in snake venom that are responsible for immobilizing prey quite likely evolved via gene duplication from pancreatic enzymes.

Long before genes had been discovered, Charles Darwin in 1859 noted in a curiously apt passage the "highly important fact that an organ originally constructed for one purpose . . . may be converted into one for a wholly different purpose." Modern evolutionary biologists have even devised a new name for such features that have so drastically shifted their functional gears: they're called *exaptations*. In contrast, traits that retain the same function for which they originally evolved are called *adaptations*. Scientists have identified many likely exaptations—animal traits that are now used in some entirely different context from their original purpose. Of course, it's tricky to determine the original purpose of some animal feature that arose several millions of years ago. Yet, as we saw earlier, good evidence exists that firefly bioluminescence originated as a warning signal to deter potential predators. Only much later, and only within certain firefly lineages, did this light-producing talent become exapted into an adult courtship signal.

For a familiar example of exaptation, consider bird feathers. Although feathers now enable modern-day birds to fly, it wasn't always so. We now know that all birds descended from theropod dinosaurs, and many different bird ancestors sporting feathers have been unearthed from fossil-rich rocks in northeastern China. Because these feathered theropods couldn't fly, their feathers must have originally provided some other advantage—perhaps they acted as fancy ornaments for courtship, or provided thermal insulation. While feathers have evolved to become an aerodynamic feature permitting "modern dinosaurs"—birds, that is—to fly, they originated through natural selection to serve a completely different purpose. So in evolutionary parlance, both feathers and luciferase represent exaptations.

And let us not ignore the costar of this bioluminescent show. Different creatures feature differ luciferins. But in every firefly species, the same luciferin costars as the molecular dance partner to luciferase. Yet many questions remain about how fireflies acquire this key light emitter. Luciferin is an unusually talented molecule, and there's still a lot we don't know about where, when, and how it gets synthesized. Clearly, we still have many puzzle pieces to fit together before we gain a complete chemical picture of firefly bioluminescence!

## Putting Fireflies to Work

Firefly light isn't just useful to fireflies. Before electricity, of course, firefly light had many uses. I've heard oldsters around the world tell stories about gathering fireflies to use at night for reading, for biking, and for walking along forest paths. But scientific discoveries about the chemistry behind firefly bioluminescence have paved the way for even wider practical applications. Fireflies' light-producing talent has provided invaluable tools for improving public health, for facilitating innovative research, and for advancing medical knowledge.

The food industry has long used fireflies' light reaction to detect good food gone bad—that is, food that's contaminated by noxious bacteria and thus unsafe for human consumption. Test kits containing firefly luciferase and luciferin are used to detect the presence of ATP, a compound that's found in all living cells; this includes any live microbes like *Salmonella* or *E. coli* that might be lurking in our food or beverages. When firefly luciferase and luciferin are added, the ATP from these microbial contaminants produces visible light. The more ATP, the brighter the light, so the intensity of luminescence even reveals how many bacteria are present. Starting in the 1960s, this bioluminescent test has been able to detect even tiny amounts of microbial contamination, using very sensitive instruments to measure light production. This firefly-inspired test provides results more quickly than previous methods; it requires mere minutes instead of the days needed to detect contaminated food by growing bacterial cultures. This handy bioluminescent ATP assay, which now employs synthetic luciferase, is still widely used to ensure food safety by detecting microbial contamination in milk, soft drinks, meat, and other commodities.

Similar methods now aid drug discovery in the pharmaceutical industry, which relies on high-throughput screening to rapidly test potential new chemotherapies for treating cancers. Tumor cells are grown in culture and then treated with different drugs. Using luminescence-based tests to measure cell viability, those drugs most effective at killing tumor cells can be quickly identified.

Since the genetic blueprint for luciferase was isolated and deciphered in the 1990s, the number of practical uses for firefly bioluminescence has skyrocketed. Many new discoveries in medicine and biotechnology have been made using *luc*, the firefly luciferase gene, as a "reporter" for the activity of other genes. In this application, researchers splice the *luc* gene together with a specific gene they want to study, and then insert this spliced DNA into living cells. Whenever the spliced DNA gets transcribed, the cells will manufacture luciferase. When luciferin is added, these cells will respond by lighting up. This technique has been used, for instance, to find out exactly when and where specific plant genes get turned on. To learn about particular genes regulating plant growth, biologists have spliced the *luc* gene into different bits of plant DNA. When plants are sprayed or fed with luciferin-containing water, the leaves will glow whenever *luc* gets turned on. This allows researchers to identify specific genes regulating plant growth at different times and locations. Such reporter genes have also provided powerful tools for studying diseases, for developing new antibiotic drugs, and for gaining new insights into many human metabolic disorders.

Fireflies have also helped develop real-time, noninvasive imaging methods to see what's happening inside living organisms. When *luc* genes are used to label particular cell or tissue types, very sensitive cameras can be used to detect their light inside the live animal. By labeling cancer cells in mice, scientists have identified new anticancer drugs that halt tumor growth and reduce the likelihood of cancer metastases. Related methods have helped find new drugs that can be used to combat tuberculosis. The bacterial pathogen responsible for this disease has evolved resistance to even our most powerful antibiotics, and thus the disease been difficult to eradicate. To help discover new treatments for antibiotic-resistant tuberculosis, scientists have infected mice with luciferase-labeled tuberculosis bacteria. They then treated the mice with various antituberculosis drugs and used bioluminescence imaging to monitor the bacteria inside.

All these advances in public health, medicine, and scientific research were made possible by scientific discoveries about the biochemistry of firefly light production. This list goes on and on. And it goes to show just how much benefit we humans can derive from evolution's natural inventiveness.

## Controlling the Flash

Deciphering how chemical energy gets converted into light proved to be merely the first step in understanding firefly light signals. How do fireflies manage to channel this chemistry into a tool for communicating with one another? This chemistry is played out on a biological stage that involves exciting nerve impulses that originate in the firefly's brain, the architecturally elegant structure known as the firefly lantern, and miniscule luminous organelles tucked deep inside cells. A great deal of our knowledge concerning how fireflies flash is based on nearly sixty years of scientific research conducted by an insect physiologist, the late John Bonner Buck. Beginning in the 1930s, John Buck and his students set up careful laboratory experiments that still form the basis for our understanding of how, and where, fireflies produce their light.

Born in 1913, John Buck attended Johns Hopkins University for both his undergraduate and doctoral studies. It was here that Buck fell under the spell of the Big Dipper firefly *Photinus pyralis*, a common denizen of his Baltimore backyard. In 1933, he decided to spend his summer vacation deciphering what cues fireflies use to begin their nightly flash activity. Earlier workers had noted that fireflies sometimes began flashing much earlier on cloudy days, which suggested they might rely on daily cues provided by diminishing light at dusk. Or do they follow an intrinsic twenty-four-hour cycle?

To test these different hypotheses, Buck converted a university darkroom into a summer camp for his backyard insect friends. He collected hundreds of male *Photinus pyralis* fireflies in glass milk bottles, brought them into his darkroom, and exposed them to different light regimes. After releasing the fireflies into a mesh cage, he counted their flashes. In one experiment, Buck discovered that the fireflies started flashing whenever he lowered the lights from bright to dim, no matter what time of day it happened to be. In another experiment, he kept bottled fireflies in continuous darkness, then released them into the cage. Camped out inside the darkroom in his sleeping bag, Buck spent four straight days counting the fireflies' flashes for five minutes every hour on the hour. Even when they'd been kept in constant darkness, these fireflies continued to flash according to an internal twenty-four-hour cycle. By counting flashes in the dark, Buck had discovered the first circadian rhythm in light production, and also showed that diminishing light intensity cued the exact onset of fireflies' flash activity.

By 1936, Buck had completed his PhD dissertation, succinctly titled "Studies on the Firefly," and soon he would elevate his backyard *Photinus pyralis* into the best-studied firefly on Earth. In 1939 he married Elisabeth Mast, the daughter of his major professor at Johns Hopkins. Elisabeth became not just his wife but also his trusted scientific companion over the next sixty-five years. Buck spent his career at the National Institutes of Health in Bethesda, where he directed the Laboratory of Physical Biology. Over the next several decades Buck's research program would delve deeply into firefly cell biology, anatomy, and neurophysiology, intently focused on understanding the mechanics of firefly flashing at the level of cells, at the level of tissues, and at the level of the whole organism. This breathtaking scope has yet to be matched by any other single researcher.

While many creatures can glow, fireflies are among a mere handful that can control their light emission to make discrete flashes. To understand how fireflies control their light so it shines out at just the right time and in just the right place, we'll need to consider some microscopic structures that lie deep inside the firefly's lantern.

## A Journey inside the Firefly Lantern

One of Buck's many contributions included bringing to light the anatomical details of firefly lanterns, which he revealed to be intricate structures matched with physiological complexity. Some fireflies merely glow, shining out their light across many seconds or minutes, while others emit flashes, quick bursts of light. These differences in bioluminescent control are reflected in the structural complexity of their respective lanterns. The most elegant internal architecture I've ever seen appears in the lanterns of flashing fireflies like adult *Photuris* and *Photinus*. Located behind a transparent layer of cuticle on the firefly's underside, the lantern extends across one or two abdominal segments. Inside what looks like a simple slab of tissue, the lantern contains about 15,000 light-producing cells called photocytes. Wedge-shaped, these photocytes are arranged in concentric rosettes that resemble orange segments sliced in cross section (Figure 6.2). This exquisitely anatomical arrangement is what allows fireflies to exercise precise control over exactly when and where their light shines.

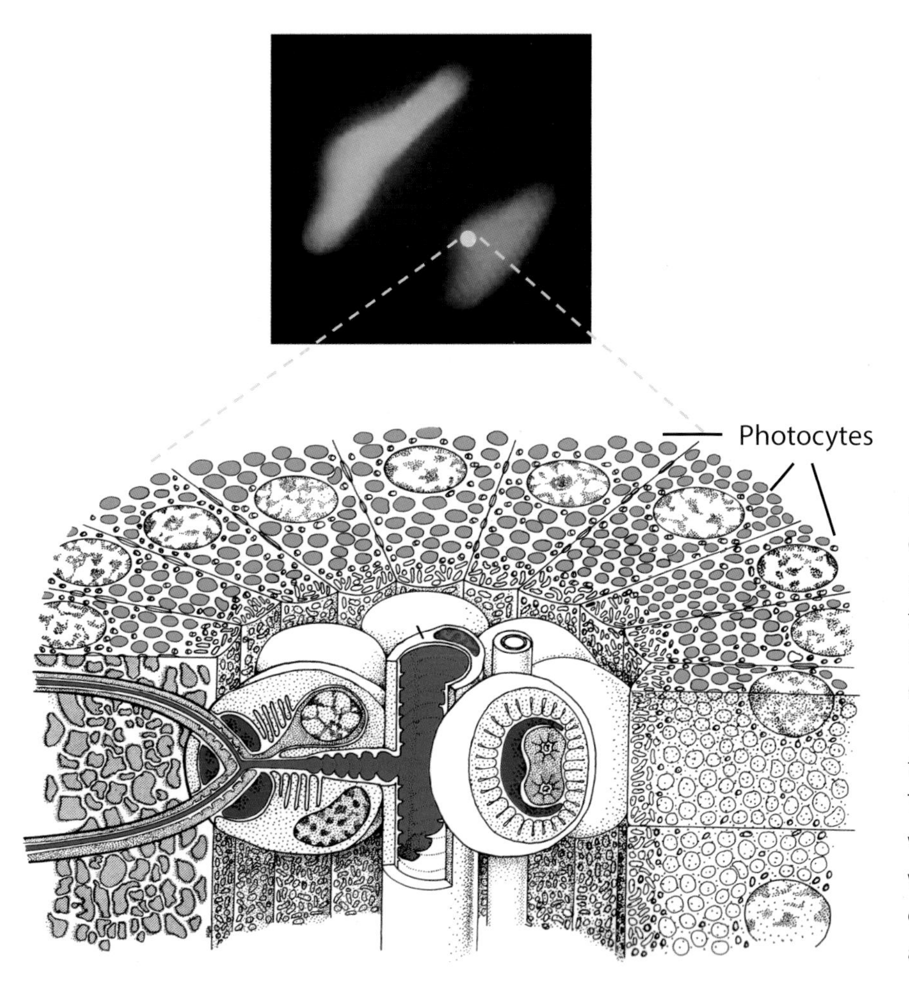


**FIGURE 6.2** A firefly's lantern contains thousands of light-producing cells, called photocytes, arranged into circular rosettes. Each rosette surrounds a central air tube (blue) with nerves (red) that terminate on nearby cells. The photocytes are packed with peroxisomes (green), which house luciferin and luciferase (adapted from Ghiradella 1998).

All the light-producing action takes place within the confines of the photocytes. Inside each photocyte the chemicals that produce light are housed within hundreds of organelles called peroxisomes; these are so numerous, they occupy nearly one-third of the cell's total volume. This is where the luciferase-luciferin complex is stored, waiting for exciting oxygen to come along to make it flash.

Like all insects, fireflies get their oxygen delivered through a network of tiny air tubes that run throughout their bodies. Within the lantern, an air tubes plunges down the center of each cylinder, branching out laterally at the level of each rosette. Tiny branches run out between the photocytes and supply the oxygen required for the final step in the light-producing reactions. Most children know that a firefly lantern smeared across your forehead will glow for hours. So researchers have long assumed that fireflies control their flash by somehow regulating how much oxygen gets to their photocytes. But anatomical studies of firefly lanterns have been unable to detect any mechanical valves that could open and shut rapidly enough to control flashing. However, while most air tubes are stiffly reinforced to remain open, lantern air tubes suddenly become flimsy and collapsible for a short stretch at their final branch point. It has been proposed that if they collapsed during daylight hours, these air tubes might help keep the lantern dark by reducing oxygen flow to the photocytes.

Yet when night falls, the flashing begins. Each flash is triggered by a nerve impulse born in the firefly's brain. Here, a pacemaker generates the rhythmic pattern that keeps each firefly flashing to the beat of its own drummer. From the brain, the nerve impulse traverses the nerve cord down to the last abdominal segment where nerve fibers transmit the electrical signal into the lantern. Within the lantern, nerve endings secrete octopamine, the insect equivalent of the human neurotransmitter adrenalin. Finally, the flash comes to fruition within the photocytes.

Several details of lantern architecture differ between those fireflies that can merely glow and those that can produce precisely timed flashes. First, in flashing fireflies the nerves carrying the signal that originates in the firefly's brain are *not* directly wired to the light-producing photocytes. Instead, the nerves terminate on cells nearby. Second, flashing fireflies keep their photocyte interior conspicuously well organized. Thousands of mitochondria, the powerhouses of the cell, are densely packed around margins adjacent to the air tubes, while the peroxisomes housing luciferase and luciferin are sequestered away in the cell's interior.

What about those fireflies that merely glow? This includes larval fireflies, all of which can produce only a slow glow that gradually rises and falls. And glow is the height of bioluminescent fashion for many firefly females. Females of the common European glow-worm *Lampyris noctiluca* can glow continuously for several hours, yet they cannot flash. Anatomically, the lanterns of these glowing fireflies are much simpler than those of their flashier cousins. In the larvae of *Photuris* fireflies, the lantern consists of two small disks, each about one-half millimeter in diameter, located on their last abdominal segment. Each lantern contains about 2,000 photocytes, but these are randomly arranged. Inside the photocytes, mitochondria and peroxisomes are all jumbled together, lacking the fastidious organization these organelles show inside flashing fireflies. Also, in glowing fireflies the nerves that stimulate the glow connect directly to each photocyte.

Form reflects function, and these small architectural differences hidden within the lantern pointed the way toward the secret switch that allows flashing fireflies to turn their lights on and off again so quickly.

## Discovering the Firefly's Light Switch

Firefly adults are among the few living creatures capable of precisely controlling their bioluminescent output. This light switch is what enabled flashing fireflies to evolve their elaborate Morse-code-like courtship signals. Until 2001, we knew little about how fireflies managed their internal chemical reactions to make the precise flashes and flickers they use for sexual communication.

A serendipitous lunchtime conversation with colleagues at Tufts University sparked a fun and productive scientific collaboration. It was springtime, so, after discussing the Red Sox, we started talking about fireflies and the inner working of their lanterns. We were puzzled by a mystery—how did the nerve signal get carried from the nerve synapse to the light-producing photocytes? Our lunch happened to include a propitious combination of experts—an insect neurobiologist, a biochemist, and an evolutionary ecologist—so we set out to answer this question. The research team also included my husband Thomas Michel, a Harvard Medical School professor who studies the biological roles of nitric oxide. This simple but important molecule—composed of merely a single nitrogen atom cojoined to one oxygen atom—is commonly abbreviated as NO. This short-lived, quickly diffusing gas is manufactured by enzymes and helps to carry messages between cells. Within the human body, NO is responsible for controlling everything from blood pressure, to penis erections, to learning and memory. A truly versatile molecule, NO also performs diverse biological functions in other animals.

We were especially intrigued by the strong effect that NO has on mitochondria; NO is known to temporarily shut down the normally active respiration of these cellular organelles. When mitochondria stop their respiration, they stop using oxygen. Could nitric oxide somehow be regulating oxygen supply to the light-producing reactants housed inside the firefly's photocytes?

My husband and I were thrilled to be working together, our first joint research project since our undergraduate days together at Harvard. This romantic project also included our two sons, then eight and eleven years old, who volunteered as field assistants to help collect fireflies (Figure 6.3). After bringing the fireflies into the lab, we placed them inside a tiny, custom-designed chamber that we alternately filled with plain air or added NO gas. To our surprise, whenever we turned on the nitric oxide, the test fireflies glowed or flashed almost continuously. And just like flipping a light switch, they turned dark once the nitric oxide was turned off. We also did some experiments in which we added certain drugs to isolated firefly lanterns bathed in a saline solution. Normally, when you add the neurotransmitter octopamine to a firefly lantern, it lights right up: firefly lantern + octopamine ⟶ light. But we found that the normal flash response to octopamine was totally blocked whenever we added chemicals known to inactivate NO. So firefly lantern + octopamine + NO inactivators ⟶ darkness.

**FIGURE 6.3** Chasing a better understanding of firefly flash control became a family affair: *Clockwise from left*: my husband Thomas Michel, me, and our sons Ben and Zack (photo courtesy of Harvard News Office).

These experiments had given us new insight into the firefly's light switch, showing that NO is involved in flash control and also that cells in the firefly lantern make nitric oxide. But how exactly might this work? We proposed the following sequence of events. When a nerve signal arrives in the lantern, the release of octopamine stimulates nearby cells to begin producing NO. This quick-diffusing gas soon reaches the adjacent photocytes, where it encounters the millions of mitochondria that are densely packed around the edges. Here, NO opens the oxygen gateway by temporarily halting mitochondrial respiration. Once the mitochondria stop respiring, the oxygen they'd normally be consuming can now freely diffuse into the interior of the photocyte. And it is here, inside the peroxisomes, that the activated luciferin-luciferase team is just waiting for a breath of fresh oxygen to come along and complete the chemical reaction. A flash of light! When the nerve signal stops, so does NO production, and all the photocyte's mitochondria can power up again. As these power plants begin respiring, they sop up all the oxygen as it enters the photocyte. Cut off from their oxygen supply, the light-producing reactions inside the peroxisomes stop and the lantern goes dark once again.

And so nitric oxide seems to be the secret ingredient in the firefly's light switch. Produced in the firefly lantern in response to a nerve impulse coming from the brain, NO flips the switch on mitochondrial respiration to open and close the gateway for incoming oxygen to reach the light-producing reactants sequestered deep inside the photocytes. We'd discovered an entirely new biological function for the ubiquitous signaling molecule, NO. NO plays a key role not only in human sex—by controlling penile erection—but also in firefly sex, by allowing flash communication. In 2001, we published these discoveries in the journal *Science*.

Later, we had an opportunity to present our results during "Fireflies @ 50," a symposium that celebrated a half-century's worth of contributions that the physiologist John Buck had made toward understanding how fireflies flash. Although he was ninety years old and nitric oxide was new to him, Buck was thrilled to hear about this new mechanism of firefly flash control.

## Getting in Sync

John Buck's interest in the control of firefly flashing didn't stop at just understanding flash control within a single firefly, but extended to understanding how flashing gets coordinated among hundreds of individual fireflies. Early in the 1960s John became fascinated by some *Pteroptyx* fireflies found along tidal rivers in Southeast Asia. Here, it had been reported, thousands of male fireflies congregate in trees and flash rhythmically, all in unison, maintaining their precise synchrony for hours each night. Although there were skeptics, a naturalist living in Thailand in 1935 had described some dazzling displays:

> Imagine a tree thirty-five to forty feet high thickly covered with small ovate leaves, apparently with a firefly on every leaf and all the fireflies flashing in perfect unison at the rate of about three times in two seconds, the tree being in complete darkness between the flashes. Imagine a dozen such trees standing close together along the river's edge with synchronously flashing fireflies on every leaf. Imagine a tenth of a mile of riverfront with an unbroken line of *Sonneratia* trees with fireflies on every leaf flashing in synchronism, the insects on the trees at the ends of the line acting in perfect unison with those between.

"Then," he concluded, "if one's imagination is sufficiently vivid, he may form some conception of this amazing spectacle." It was also rumored that such fireflies congregated in the same trees night after night, for months, enabling native boatmen to use firefly display trees as navigational beacons when they paddled along the waterways at night.

John was hooked. In 1965 he and Elisabeth embarked on a trip to Thailand funded by the National Geographic Society. They headed for the Mae Klong River south of Bangkok, where they hired a local water taxi at dusk. They nosed in among the tangled roots of the *Sonneratia* mangrove trees to watch the motionless males all flashing in unison like Christmas tree lights. From the gently rocking canoe, they captured the first scientific documentation of firefly flash synchrony using a light-recording photometer and a 16 mm film camera. They also captured some of these *Pteroptyx malaccae* fireflies—as Elisabeth would later recall, "You could just reach up and shake the branches, and fireflies would rain down." When they later released their captives back in their Bangkok hotel room, the fireflies flew around briefly and then settled down on the walls and furniture. The Bucks watched as the males began flashing, at first synchronizing in small groups. Soon, the entire roomful blossomed into synchrony.

For the Bucks, and likely for everyone who sees it, the pulsating, hypnotic display of synchronous fireflies was a life-altering experience. Other creatures, including frogs, crickets, and cicadas, gather in groups where they can sometimes produce an impressive chorus of synchronized croaks, chirps, or clicks. Yet the visual synchrony of flashing fireflies—thousands of sparks powerfully pulsating in total silence—is unforgettable.

The Bucks eventually published their findings in the journal *Science*; after eight pages of technical analysis they dryly stated: "Our records show that synchronous flashing in Thai fireflies is not an illusion." Over the next fifty years John Buck and his students would record, measure, and experiment to decipher the physiological mechanisms that allow these and other fireflies to maintain their massive flash synchrony.

Such synchronous fireflies, like the myriad pacemaker cells that control the beating of the human heart, embody a mathematical concept known as "pulse-coupled oscillators." Lucidly explained by Steve Strogatz in his book *Sync*, such systems consist of many distinct entities ("oscillators"), each controlled by its own internal metronome. As each firefly's metronome blinks out its own rhythm, its timing automatically gets adjusted in response to incoming flashes (hence "coupled"). Each evening when firefly males begin their courtship display, their flashes are initially sporadic and uncoordinated. But as males adjust their personal flash rhythms to match up with what they see coming from their neighbors, synchrony emerges spontaneously.

Buck focused his studies on learning how fireflies reset their internal metronomes. He discovered that in some species, this adjustment happens by a "phase delay" mechanism. Here, a male firefly either shortens or lengthens a single flash cycle depending on what he sees happening around him. This one-time adjustment is followed immediately by a return to the male's normal flash rhythm. In other synchronizers, like the *Pteroptyx malaccae* fireflies that maintain their precise synchrony over several hours, the mechanism generating group synchrony is more complex. In such species, the males continually readjust their own internal rhythm, flashing faster or slower to match what's happening around them.

Thanks to the work of John Buck and his students, we now have a pretty good understanding of *how* some fireflies manage to synchronize their flashes. Yet exactly *why* males of certain firefly species have evolved this ability to synchronize remains a mystery. Indeed, attempts to explain the evolutionary forces responsible for firefly synchrony managed to kick off an acrimonious feud that divided firefly biologists for several decades.

# Science Confidential

On a beautiful, breezy New England day during the summer of 1985, I found myself zipping across Woods Hole harbor with John Buck and his wife, Elisabeth, on their sailboat, an 18-foot Cape Cod Knockabout. Stately and keen-eyed, John had an old-school gentlemanly manner that overlaid a quiet determination. Somehow I had the tiller and was anxiously trying to avoid what I was convinced would be a fatal collision in this chaotic harbor. I needn't have worried. Having summered in Woods Hole for many years, the Bucks were regulars in the yacht club's weekly Knockabout races, and John even chronicled these weekly racing adventures in the local newspaper under the byline "Old Salt." Back from sailing, we spent many hours talking fireflies in the welcoming warmth of their Woods Hole summer home.

I'd traveled to Woods Hole that day mainly for the honor of meeting this towering figure who'd contributed so much to our knowledge of firefly biology. I also was hoping to gain some insight into why this Quaker pacifist had been embroiled for so long in a bitter feud with Jim Lloyd, the field biologist we met in chapter 3. When I began studying fireflies in the early 1980s, I quickly learned that firefly science in the United States was sharply divided into two warring factions. On one side stood Jim Lloyd and his students, on the other side John Buck and his students; the two sides were definitely not on speaking terms. I was young, I was not formally affiliated with either side, and I had interesting questions about sexual selection in fireflies that I was trying my best to answer. Luckily, I was able to maintain good relations all around. But this long-standing scientific feud had been simmering for decades, and I'd heard many horror stories about how it had ruined others' careers.

What was it all about? From my discussions across both sides of this aisle, I learned there had been a row in the mid-1970s concerning unpleasant peer reviews of manuscripts that each group had submitted for publication. But this happens routinely in science, and such mundane disputes can't explain a feud that had lasted several academic generations. Or perhaps they disagreed about where to best conduct their firefly research. Buck's modus operandi across his many decades of active research was to conduct his experiments under carefully regulated conditions that are only feasible in a laboratory. It's a fact of scientific life that achieving such tight experimental control means sacrificing knowledge about how the animal behaves in its natural setting. On the other hand, Jim Lloyd spent a lifetime learning about fireflies by watching and recording their behavior outdoors in the wild. Yet Buck certainly did his share of fieldwork. He'd worked with *Photinus* fireflies in his Baltimore backyard and around Woods Hole, and he headed several expeditions to Southeast Asia to study synchronous fireflies. So a simple split between lab- versus field-based biology seems unlikely.

Instead, these two firefly researchers locked horns over an even more fundamental difference in their scientific perspectives. One man spent his career intently focused on asking "how" questions, while the other focused on asking "why" questions. In a famous paper published in 1963, Niko Tinbergen paved the way for a more integrative approach to the scientific study of animal behavior. He presented four questions that any curious scientist might ask when trying to explain some feature of an animal's behavior, or morphology, or physiology: (1) What is this trait used for? (2) How did it evolve? (3) How does it work? and (4) How does it develop during this animal's lifespan? The first two have been called "ultimate" questions because they ask *why* animals exhibit particular features. The central goal in answering these questions is to understand how such features evolved, and how they currently impact an organism's ability to survive or to reproduce. This is the goal of behavioral ecology. The last two are often called "proximate" questions because they focus on *how* an animal's features work, and their central goal is to build a mechanistic understanding of these features. This is the purview of animal physiology.

Beginning in the mid-1960s and continuing through the late 1980s, this ultimate-proximate tension led to several skirmishes that flared up around various questions in firefly biology. But nowhere did the argument get more heated than around possible explanations for firefly flash synchrony. Answering ultimate questions requires a sophisticated understanding of evolutionary theory, so biologists who concentrate on proximate mechanisms sometimes shy away from asking questions about ultimate causation. Scientists are people, and people have different proclivities. John Buck had been trained as a physiologist and, as we've seen, he was the leading expert on the proximate mechanisms that permit synchrony. But my conversations and correspondence with Buck convinced me that the evolutionary thinking necessary to answer ultimate questions was foreign to him. Lloyd, meanwhile, was trained as a behavioral ecologist and interested in the ultimate question of *why* fireflies synchronize their flashes.

In chapter 2 I described some hypotheses scientists have proposed to explain firefly synchrony. We still lack a definitive answer, but John Buck and Jim Lloyd expended quite a lot of energy arguing about these ideas. They vehemently critiqued each other's explanations for what evolutionary benefits these synchronizing males might enjoy. Buck maintained that synchrony could evolve if it provided some kind of advantage to the entire group of males. But Jim Lloyd pointed out that selection acts most strongly on individuals, so synchrony must also somehow enhance the reproductive chances for each participating male. All the neural machinery and other traits that make certain male fireflies act as pulse-coupled oscillators will endure only if they provide a reproductive advantage for individuals, in addition to any group benefit that synchrony might provide.

We still don't know exactly what these advantages might be—and they're likely to differ between traveling synchronizers like *Photinus carolinus* and stationary synchronizers like *Pteroptyx* fireflies. Sadly, Buck and Lloyd never managed to reconcile their differences. Although both men devoted their careers to deepening our understanding of firefly biology, their ships collided in the scientific night—caught in fog, they were blinded by their disparate approaches to posing biological questions.

But we've learned that synchrony, a paradoxically cooperative male behavior, is merely the first stage in these fireflies' courtship rituals; once a female appears on the scene, males' cooperation abruptly ends. Now it's every man for himself as they gear up for competition mode, each male trying to stand out from his rivals. Lynn Faust has described what happens when males of *Photinus carolinus* finally spot a female's reply. They drop out of synchrony and switch from six-pulsed to single-pulsed flashes. As they approach the female, males begin what Faust describes as "chaotic" flashing. Rival males alternately give off rapid-fire bursts of light, illuminating the night like firecrackers. Turning competitive, males crowd around the female, grappling fiercely and shoving one another with their head shields. Even after one victorious male mates successfully with the female, his ever-hopeful rivals remain piled atop the mating pair for hours.

We know much less about what happens once a *Pteroptyx* female flies into a display tree filled with synchronizing males. Working on *Pteroptyx tener*, one researcher reported that when a female landed nearby, a male would twist his abdomen around nearly 180 degrees to flash directly in her face. Perhaps this lets the female assess the male's brightness, or perhaps he aims to blind her to other males' advances. In other *Pteroptyx* species, males seem to use some yet-to-be-identified chemical signal to impress females. We're still mostly in the dark about how male-male competition and female choice play out in synchronous fireflies, and many questions remain to be answered about firefly synchrony.

* * *

A steel gray filing cabinet in my office holds two large manila folders, one labeled Buck, the other labeled Lloyd. These folders are filled beyond bursting with all the scientific articles each man published over his long and highly productive career. Printed on glossy paper and personally signed by the authors, these reprints represent a venerable academic tradition that's been extinguished by the advent of the pdf and online scientific publication. Still, I hold onto these folders because their heft so tangibly captures the sheer magnitude of each man's contributions to firefly biology. And I find myself wondering how much more we might have learned if these two scientists had recognized that their proximate versus ultimate scientific perspectives were actually complementary rather than antagonistic. While their disagreements sprang from a universal dichotomy in biological research, their scientific feud cast a long shadow over firefly science. The worst collateral damage from this feud was probably all the promising young students who were deterred from studying fireflies by the clash between these titans. Fortunately, this discouraging pall is beginning to fade as a new generation of young scientists is springing up around the world, now working together to expand our understanding of firefly biology.

In this chapter, we've taken a deep dive inside the firefly lantern to discover how fireflies make their magical lights, and to explore how this remarkable light-producing ability might have evolved. Now, however, it's time to turn away from the light and to confront the dark side of fireflies.

CHAPTER 7

* * *

# Poisonous Attractions

*Never came poison from so sweet a place.*

- William Shakespeare -

## For the Love of Insects

I'll wager that no one has ever sniffed and nibbled as many bugs as the late Tom Eisner, an entomologist and Cornell professor for more than fifty years. Eisner's path to fame as the world's leading chemical ecologist led across many countries and continents. Eisner early on became enchanted with insects, a portable fascination that conveniently matched his peripatetic childhood. When Eisner was three, his family left Germany to escape Hitler. They moved to Barcelona, where they ran smack into the chaos of the Spanish Civil War. The family finally landed in Uruguay, where the teenaged Eisner could spend blissful hours wandering outdoors, searching under rocks and logs for the bugs he loved. South America ravished him with its bounteous biodiversity. Young Eisner's fondness for insects was also nurtured by his father, a pharmaceutical chemist who enjoyed perfume making as a hobby. Perhaps these subtle fragrances helped hone Tom's trademark talent, his keen sense of smell. In any case, Tom Eisner would eventually fashion an illustrious career that combined insects, chemistry, and behavior to probe the secret strategies behind insects' remarkable evolutionary success.

In 1957 Eisner arrived at Cornell, where he soon struck up a partnership with the chemist Jerrold Meinwald to find out how insects use chemicals to defend themselves against their many enemies. Working together, Eisner and Meinwald created the new discipline called chemical ecology, now a thriving research field. A self-described biologist-explorer, Eisner always retained his keen interest in natural history. On long walks he'd immerse himself in the Lilliputian world of insects and their kin, closely observing their encounters with predators and recording their intriguing habits. Especially attracted by the pungent odors some insects give off when handled, Eisner learned the best approach was to sniff insects cautiously. Inspired by his field observations, he and Meinwald would then embark on detailed laboratory experiments and chemical analyses. During their more than fifty years of highly productive collaboration, these two scientists discovered that insects use a dizzying array of schemes to fend off adversaries: acid jets, sticky secretions, waxy hideouts, paralytic liquids, repellent mists, and boiling caustic sprays. Forged to ensure survival in a hostile world, such varied weaponry highlights insects' chemical ingenuity and splendidly demonstrates the creative power of evolution.

With a knack for scientific storytelling, Eisner shared his discoveries not only through his 500+ scholarly articles but also through popular books, films, and interviews, always with delight twinkling in his eye or flowing from his pen. Eisner was also a talented photographer, filling his scientific papers and books with spectacular action shots of insects defending themselves. Throughout his long career he unabashedly proclaimed his admiration for insects, explaining simply "Once you fall in love with them, you can't fall out of love."

## Lightningbugs for Breakfast? No-No!

It was while studying under Tom Eisner in the mid-1960s that Jim Lloyd, the biologist who deciphered fireflies' flash codes, earned his PhD degree. So it was probably inevitable that Eisner would eventually turn his powerful scientific gaze toward fireflies. We've seen that fireflies congregate for courtship, where they converse with potential partners using highly conspicuous, bright flashes of light. Surrounded by bats, toads, and other hungry insect eaters, how do these creatures manage to advertise their sex appeal so brazenly?

Eisner's explorations into fireflies' chemical arsenal began in the mid-1970s. One summer he and his family undertook a joint project in which they enlisted the help of a feathered collaborator: his pet bird, a Swainson's thrush named Phogel. Like Eisner, Phogel also loved insects, but the bird enjoyed them for breakfast. Each morning, Eisner walked out early to collect whatever insects he could find. After finishing their own breakfast, the Eisner family sat back to watch the action. The insects were tipped, one by one, out of their vials and into Phogel's feeding tray. Phogel turned out to be quite a finicky gourmet. He enthusiastically downed most insects, which the family scored as *yum-yums*. Other insects he pecked once, then rejected forevermore. Scored as *no-nos*, these earned sufficient disdain that Phogel still remembered them two weeks later when Eisner presented them again. The remaining insects fell into an intermediate *so-so* category; they were pecked, then eaten or not depending on how hungry Phogel was. That summer, Phogel dutifully passed judgment on the tastiness of some five hundred representatives from more than one hundred different species. Fireflies, it turned out, were among the very few insects that consistently earned a definitive *no-no* judgment from this bird.

Phogel wasn't alone in his disdain for fireflies. Quite a few creatures that normally make a living eating insects find fireflies repugnant. Jim Lloyd compiled a slew of such anecdotes, reporting similar firefly aversions in monkeys, toads, lizards, geckos, chickens, and various other birds. When Lloyd tried offering *Photinus* fireflies to an *Anolis* lizard, it quickly snatched the prey but spat it back out just as quickly. Following this encounter, the lizard spent several minutes vigorously wiping off its snout. Just like Phogel, lizards apparently find firefly encounters unpleasant, and sufficiently memorable that they'll avoid these insects for weeks.

Some other lizards were not so smart. In the late 1990s stories began circulating in veterinary circles concerning unexplained deaths of *Pogona* lizards, exotic reptiles commonly known as bearded dragons. These popular pets are now bred in the United States, but they were originally imported from Australia. The reason for their demise was discovered by a vet who learned that some well-intentioned owners had given their pet lizards some locally collected fireflies. The bearded dragons readily swallowed these insects, but soon began violently shaking their heads and repeatedly gaping their mouths. Their normally tan bodies turned black, then they keeled over and died. I've never met one, but bearded dragons apparently lack common sense enough to forgo fireflies, perhaps because such toxic insects are rare in their native habitat.

Birds and lizards clearly eschew fireflies, even though these day-foraging insectivores might often encounter these insects resting on plants. But what about nocturnal insectivores like bats, toads, and mice? In New England, researchers collected the fecal pellets deposited by four different bat species. No fireflies showed up in the diets of 260 bats, even though all were netted at sites with high firefly activity. Bats that were tested in captivity readily ate mealworms (these are *Tenebrio* beetle larvae commonly sold as food for pet birds, reptiles, and other insectivores). Yet when the researchers painted the mealworms with a solution made from ground-up *Photinus* fireflies, the same bats found them decidedly unappetizing. Even a quick lick made the bats cough, shake their heads, and vigorously wipe their snouts. When toads and mice were similarly tested, they too avoided firefly-coated mealworms.

So a plethora of evidence proves that many otherwise voracious insect eaters fastidiously avoid making a meal from fireflies. Even Tom Eisner, who had quite a reputation for taste-testing insects, gave up sampling fireflies because they tasted so terrible. The lesson is clear: you can admire fireflies all you want, but don't eat them!

## Chemical Weapons

Back in Ithaca, it was Phogel's discerning palate that first inspired Eisner and his Cornell colleagues to analyze fireflies' defensive arsenal. They were curious: what exactly was it that made fireflies so repugnant to this pet bird? Following a decades-long quest, they'd eventually uncover a tale brimming with enough poison, passion, subterfuge, and death to rival even the most exciting spy thriller.

They first enlisted five wild-caught hermit thrushes as additional taste-testers in laboratory trials to see if these birds shared Phogel's low opinion of fireflies. In these trials, each hermit thrush was sequentially offered sixteen food items. Presented in random order, one-third of the offerings were *Photinus* fireflies, while the rest were tasty mealworms. The birds voted with their beaks, and the verdict came in. The hermit thrushes gobbled up 100% of 274 mealworms. Yet they ate just one of the 135 fireflies proffered; the unfortunate bird who'd eaten the one firefly soon vomited it up. Clearly, fireflies taste revolting to these normally insectivorous birds. The scientists also noticed that, even after getting pecked, many fireflies walked away unscathed by their avian encounters.

To identify exactly what made these fireflies so repellent, Eisner's team collected some fireflies and extracted their chemicals. The researchers discovered that firefly blood carries a potent cocktail of bitter-tasting and toxic steroidal

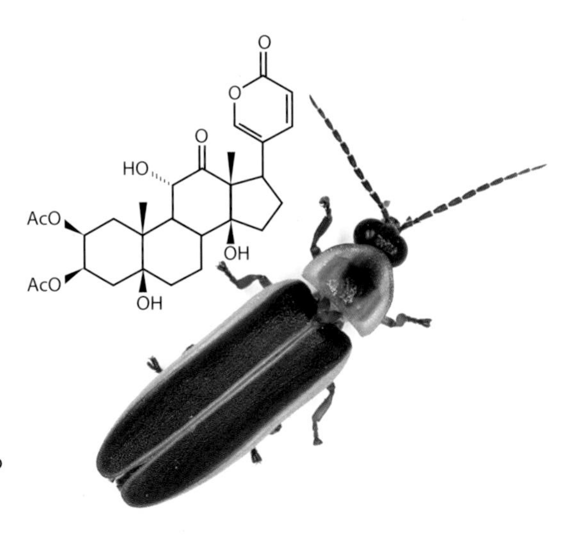


**FIGURE 7.1** Firefly chemical weaponry: *Photinus* fireflies defend themselves against most predators using toxic steroids known as lucibufagins (photo by Patrick Coin).

chemicals (Figure 7.1). They named these toxic steroids *lucibufagins*, combining the Latin *lucifer* (light bearer), with *Bufo*, after the toads that produce similar chemicals.

Yet fireflies are brimming with chemicals—were these lucibufagins specifically responsible for preventing attacks? Once again, hermit thrushes were enlisted for a bioassay designed to answer this question. Mealworms were again on the menu, but this time half these tasty treats had been coated with lucibufagins extracted from fireflies. The birds consumed 93% of the untreated mealworms, but with lucibufagin-coated mealworms their consumption fell to 48%. Coating mealworms with firefly lucibufagins definitely made them less palatable to predators.

Nature turns out to be a remarkably inventive chemist. Eisner and his colleagues discovered that a single firefly species can manufacture many distinct, yet chemically related, lucibufagins. They do so by embellishing the same chemical skeleton with various molecular groups. All these firefly lucibufagins belong to a larger class of toxic steroids known as the bufadienolides (again, named for toxins found in the skin of *Bufo* toads). These compounds make great poisons because they're effective against almost any animal. At high doses, bufadienolides disable an enzyme that's vital for all animal cells. Known as the sodium-potassium pump, this enzyme actively transports electrically charged elements—sodium and potassium ions—across the cell membrane. This generates an electrical potential that enables animals to do really important things like think and move their muscles. As a result, many plants and a few animals have convergently evolved the ability to make these toxic bufadienolides as part of their chemical arsenal.

Paradoxically, many "toxic" steroids turn out to be valuable therapeutics for treating human diseases. Bufadienolides are closely related to cardiotonic steroids like the heart drug digitalis. Manufactured by the foxglove plant, digitalis is also part of nature's chemical arsenal. Ingested in low doses, these toxic steroids can have beneficial effects, as the millions of patients who've taken digitalis for their heart disease can attest. Digitalis and related compounds effectively relieve the symptoms of heart failure by strengthening heart muscle contractions and slowing the heart rate. Bufadienolides are commonly used in traditional medicines in India, South Africa, and China, where they are used to treat many problems, including infection, inflammation, rheumatism, and heart and nervous system disorders. The traditional Chinese drug Chan su (蟾蜍), for instance, is commonly prescribed for sore throats, and also as a treatment for heart failure: its major active ingredients are toad-derived bufadienolides. These compounds have also begun to attract attention as novel therapeutic agents to treat cancers that are otherwise resistant to chemotherapy. Although firefly lucibufagins haven't yet been tested, several other bufadienolides have been shown to kill cancer cells and to inhibit the growth of human-derived liver and cervical tumors in mice. Clearly, we've glimpsed only a tiny fraction of fireflies' chemical wealth, and we still have much to learn about how such chemicals might improve these insects' survival—and our own.

## A Multifaceted Defense Strategy

Like any good defense strategist, fireflies have evolved multiple tactics to fend off their enemies. Along with being toxic, fireflies also smell bad and taste terrible. These features are beneficial because they might convince a health-conscious predator to break off its initial attack, hopefully before it has inflicted serious damage on its prey.

When they're attacked, some fireflies respond immediately by exuding droplets of blood from certain parts of their bodies, a behavior called reflex bleeding. (Unlike vertebrates, where blood circulates throughout the body within arteries and veins, insects' blood circulates more freely within the body cavity.) Oozing from microscopic structures designed to rupture under pressure, this blood coagulates into a sticky glue. When an ant dares to attack a firefly, it's quickly engulfed in a bloodbath. The exuded blood thickens to coat the ant's jaws and en-

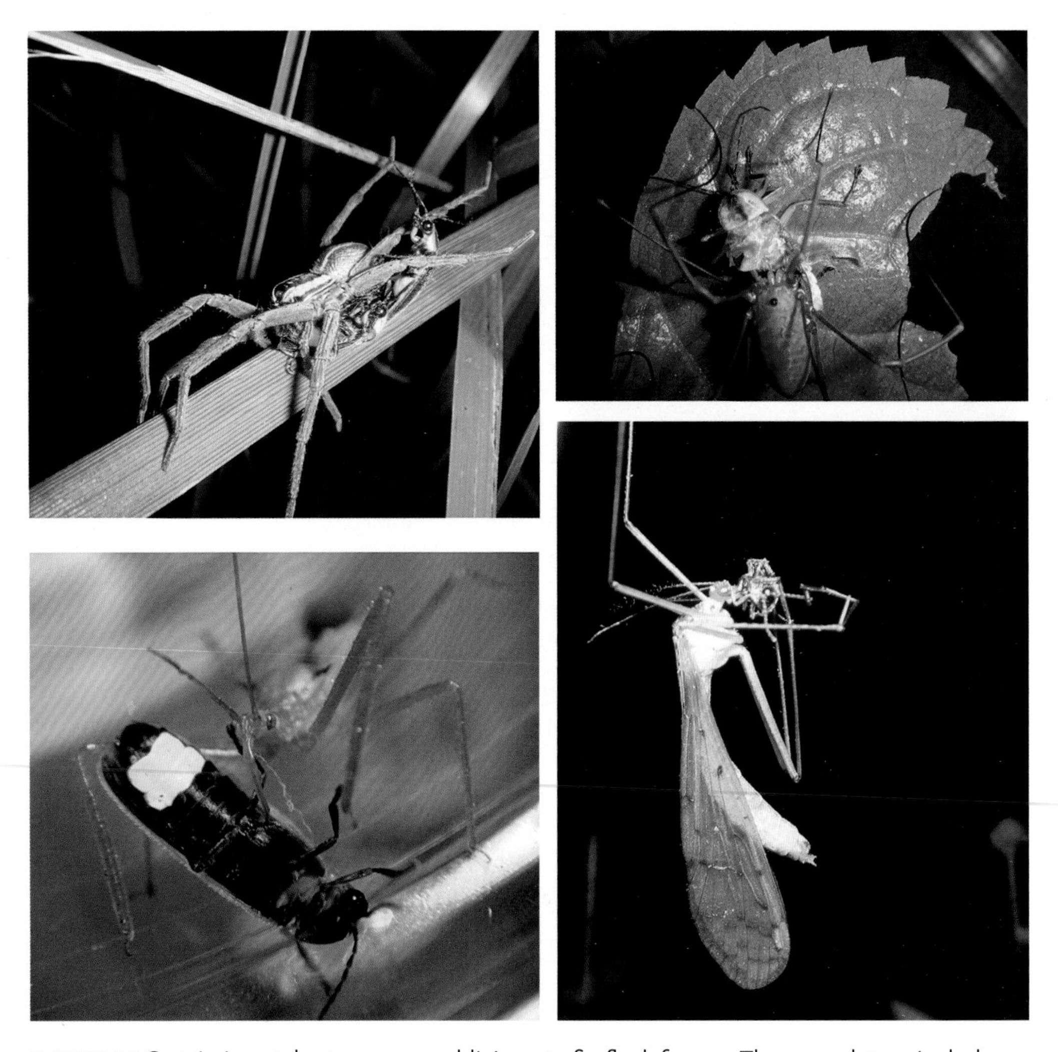

**FIGURE 7.2** Certain invertebrates appear oblivious to firefly defenses. These predators include: *Clockwise from upper left*: a wolf spider, a harvestman, a hanging fly, and an assassin bug (photos by Raphaël De Cock).

tangle its legs, immobilizing the attacker long enough for the firefly to escape. Reflex bleeding seems to be highly effective against small adversaries. But how do fireflies manage to walk away unharmed after being attacked by larger predators like birds? Reflex bleeding may play a role here, too. These heftier adversaries could be deterred not by stickiness, but by the lucibufagins carried in the blood; their bitter taste may prevent ingestion and alert attackers to the toxic consequences of eating this particular prey.

Many fireflies also repel predators by giving off a distinctive odor when they're disturbed. If you've ever mistreated a firefly, you might recall this pungent smell—it's been described as something between charred bones and that new car smell. In adult fireflies, this odor likely comes from reflex bleeding, whereas in larvae such odors come from specialized defense glands. Some larvae have pairs of tiny pop-up glands running down both sides of their body. Although the glands are normally retracted, the firefly larva quickly everts them when it's harassed by a predator. The glands release a volatile chemical that smells like pine oil or mint, depending on the species. In *Aquatica leii*, a firefly from mainland China, these larval defense glands have been shown to effectively repel attacks by fish, ants, and other predators.

In spite of fireflies' comprehensive strategy, certain invertebrate predators have somehow managed to overcome all these defenses. In 2011 I set out with colleagues for the Great Smokies, hoping to find out what predators might be exploiting the annual Light Show, where multitudes of *Photinus carolinus* fireflies gather for an exuberant mating display. Our nightly searches revealed that many insect eaters were crashing this party to feast on a luminous banquet. Among the unwanted guests were daddy longlegs, assassin bugs, and various spiders (Figure 7.2). All these predators are apparently able to circumvent fireflies' chemical defenses, though we still don't know how. And a bit further along we'll meet some other predators—females with a very peculiar penchant for fireflies.

## Evolution of Warning Displays

Fireflies have a few additional tricks to round out their comprehensive antipredator strategy. As a first line of defense they've evolved early warning signals, features that help prevent an attack even *before* it happens.

Charles Darwin's contemporary and fellow discoverer of natural selection, Alfred Russel Wallace, spent twelve years as a naturalist and collector journeying through tropical Asia and South America. In his tropical fieldwork Wallace encountered many astounding insects, including butterflies that were brilliantly colored in both their adult and caterpillar stages. Because these creatures seemed so conspicuous, Wallace wondered how they could avoid getting eaten. After studying them, he described these butterflies as being

> exceedingly beautiful and varied in their colours; spots and patches of yellow red or pure white upon a black, blue, or brown ground . . . all fly slowly and weakly; yet although they are so conspicuous, and could certainly be caught by insectivorous birds more easily than almost any other insect . . . they are not so persecuted. . . . These beautiful insects possess, however, a strong pungent semi-aromatic or medicinal odour, which seems to pervade all the juices of their system. . . . Here we have probably the cause of their immunity from attack, since there is a great deal of evidence to show that certain insects are so disgusting to birds that they will under no circumstances touch them.

When Wallace explained to Darwin his ideas about such warning coloration, Darwin wrote back that he had "never heard anything more ingenious." Wallace's insight was that merely being poisonous or tasting bad wouldn't be much benefit if an insect still gets eaten or fatally injured by a predator. To increase survival, many animals supplement their poison with a warning sign that deters potential predators *before* they attack. This "danger-flag," as Wallace called it, might be a conspicuous color pattern, or a distinctive odor, or a startling sound. Noxious animals of all shapes and sizes—from poison dart frogs to monarch butterflies to fireflies—show striking combinations of bright red, deep black, and brilliant yellow markings. These colors act as a highly visible warning, shouting out "I'm toxic, stay away!" to any would-be predators. Scientists now refer to such danger-flags as *aposematic displays*. Crafted by natural selection, such displays boldly advertise the prey's nastiness and warn predators to skip "unprofitable" prey—that is, prey whose cost outweighs its nutritional benefit. Warning displays are good for predators, too, because such features help predators to more easily recognize, and thus avoid, noxious prey.

Fireflies have mastered this early warning system; they employ both conspicuous color patterns and bioluminescence to reduce the likelihood of predator attacks. As the photographs throughout this book illustrate, many fireflies—both adults and larvae—are brightly colored, typically sporting some combination of yellow or red markings against a black or brown background. Such conspicuous coloration serves as a warning display for predators that hunt during the day, such as birds, reptiles, and some mammals. When Eisner and his colleagues offered adult *Photinus* fireflies to hermit thrushes, they noticed that after a bird had pecked and rejected a firefly once, it would subsequently avoid these noxious prey. One bad experience was enough to establish a long-lasting aversion. The thrushes learned to recognize fireflies visually, presumably using their color patterns as an early warning sign.

Larvae of the European glow-worm *Lampyris noctiluca* also wear bold colors: a jet black background sets off the bright orange spots running down their flanks. Like *Photinus* adults, these larvae also produce lucibufagins. Raphaël De Cock, whom we met in chapter 5, tested whether starlings (*Sturnus vulgaris)* could learn to avoid glow-worm larvae. Like Eisner's thrushes, when the starlings were offered mealworms and glow-worm larvae in alternation, the birds avidly consumed 98% of the mealworms, yet ate none of the glow-worm larvae. Again, the starlings at first attacked and then rejected the glow-worms, but in later trials they avoided the glow-worms on sight. These birds, too, had quickly learned to avoid this brightly colored, toxic prey.

As we learned earlier, it's thought that fireflies' ability to produce light first evolved in the juvenile stage, where it functioned as a warning display to fend off nocturnal predators. Flashing lights in the dark certainly *would* be conspicuous. But could they really improve a predator's ability to remember unpleasant encounters with toxic prey? Studies with toads and mice, predators that mainly forage at night, provide some very convincing support for this idea. After toads (*Bufo bufo*) had encountered luminescent glow-worm larvae, they became more reluctant to attack glowing artificial prey. Mice (*Mus musculus*) learned more quickly how to avoid distasteful prey (in this case, crisped rice soaked in a bitter-tasting solution) when the prey was accompanied by a flashing LED light. In both of these experiments the researchers decided to use artificial prey rather than live firefly larvae; this design let them detect whether the predators' ability to learn was due to the light signal alone—absent the larvae's warning odors.

Juvenile fireflies seem to make effective use of bioluminescence for protection, but what about adults? We've seen that in adult fireflies, evolution has shaped this biochemical talent for producing light into sexual signals used to communicate with potential mating partners. But can their flashes also ward off attackers?

I once kept a pet jumping spider, which I always suspected was a lot smarter than most people thought. A few years ago, we did some experiments where we gave *Phidippus* jumping spiders some nonluminescent, diurnal fireflies (*Ellychnia*); we already knew that these particular spiders found these fireflies distasteful. We wanted to see if the spiders would learn more quickly to avoid an unpleasant firefly encounter when we simultaneously showed them a small glowing light. And they did. When each spider's attack was accompanied by the light, it stopped attacking the distasteful prey more quickly than did spiders that hadn't seen the light. My faith in them was vindicated—though mere invertebrates, these jumping spiders successfully learned to use our warning light to avoid the nasty taste of firefly.

Like spiders, bats are also important nocturnal predators of flying insects. We've already learned that bats don't like how fireflies taste, but could bats also use flashes as an early warning to avoid a bad-tasting meal? To see whether bats avoid flashing prey, researchers made flying lures designed to imitate fireflies in flight. They found that big brown bats were less likely to attack flashing versus nonflashing lures, though flashing didn't deter two other kinds of bat. So at least some bats can use firefly flashes to identify and avoid distasteful prey.

These experiments clearly demonstrate the protective power of glow. Fireflies shine their lights not just to locate love but also to trumpet their toxicity. It turns out that other soft-bodied beetles—fireflies' closest cousins—also are chemically well defended. So it's likely that firefly toxins were already on hand before they evolved their bioluminescent warnings. Many predators learn to recognize and remember such conspicuous signals because it helps them avoid poisonous prey. Yet warning displays are more likely to make a lasting impression when they stimulate multiple senses—not just a predator's vision, but also its sense of smell and taste. As we've seen, fireflies have these bases covered, too. They've got bright contrasting colors, distinctive odors, and an unpleasant taste. All these features combine to make fireflies a thoroughly repulsive, highly memorable package with a deterrent effect that is greater than the sum of its parts.

## Firefly Look-Alikes: Tasty or Toxic?

If a creature is holding up a warning sign, then it's poisonous—right? The struggle to survive is a powerful force, so the firefly defense story merits a messy subplot. Other creatures slyly imitate the carefully crafted displays of fireflies and other well-defended creatures. The best advertisements spawn copycats; sometimes these copycats undermine the warning message, yet other times they reinforce it.

If you were to peruse any insect collection—or even click through firefly photos tagged on Flickr—you'd notice that other insects often closely resemble fireflies (Figure 7.3). Such firefly mimics have independently evolved across many unrelated groups—moths, cockroaches, soldier beetles, net-winged beetles, and

**FIGURE 7.3** Firefly look-alikes. Many distantly related insects show a surprising resemblance to fireflies: *Clockwise from upper left*: a cockroach, a blister beetle, a longhorn beetle, a net-winged beetle, a moth, and a soldier beetle (see chapter notes for photo credits).

longhorn beetles, for instance. These aren't close relatives, and so there's no reason for these other insects to show such strikingly similar color patterns. Instead, these surprising look-alikes arose through convergent evolution among unrelated species, each driven by the shared exigency of avoiding attack. This evolutionary convergence on a common danger-flag magnificently demonstrates Wallace's hunch about the potent survival value of warning displays.

Imitation may be the sincerest form of flattery, but fireflies aren't alone in attracting look-alikes. Two distinct evolutionary pathways can lead to similar mimicry complexes. The wonders of animal mimicry were first unveiled in the late nineteenth century by two European naturalists living in Brazil, Henry Walter Bates (1825–1892) and Fritz Müller (1821–1897). Both men were captivated by the sheer biotic exuberance of the tropics, where every ecological niche teems with diverse forms, all engaged in a mutual struggle for survival. Although they worked independently, Bates and Müller both spent time watching butterflies in the depths of the Brazilian rain forest. There, they would earn lasting scientific fame by revealing two new kinds of evolutionary adaptation. Now called Batesian and Müllerian mimicry, these adaptations originate through natural selection, a process that at the time had only recently been described.

Bates was an enthusiastic young entomologist specializing in British beetles and butterflies when he first met Alfred Russel Wallace. Both in their early twenties, the two men shared common interests and become fast friends, enjoying beetle-collecting trips into the countryside. Inspired by the travel adventures related in Darwin's 1839 book *The Voyage of the Beagle*, they sailed together to Brazil in 1848. There, they supported themselves by selling their specimens to amateur collectors back home. Wallace spent four years in Brazil before returning to England and heading off to the Malay Archipelago, while Bates roamed through Amazonia for several more years. During this time, he shipped about 14,000 insect specimens back to England, more than half of which turned out to be new discoveries for Western science. Much later, Bates would recall the time he spent in the Amazon as the best eleven years of his life.

It was while Bates was watching some flocks of brightly colored butterflies fluttering slowly through the forest understory that he noticed something unusual: the butterflies all appeared quite similar, in both their coloration and their leisurely flight behavior. At first glance, Bates felt certain they were all members of a single species. Yet when he inspected them more closely, he realized the flock contained butterflies from several different families; though members of the flock appeared nearly identical, they were only distant relatives. Describing this surprising phenomenon, Bates would later write:

> These imitative resemblances, of which hundreds of instances could be cited . . . fill us with the greater astonishment the closer we investigate them . . . some show a minute and palpably intentional likeness which is perfectly staggering.

Bates realized that some tasty butterflies within the flock were getting a predator-free ride by mimicking the warning colors of a more abundant, noxious-tasting species. In 1862 the biologist penned an essay describing how natural selection might have created such mimics. First, as Wallace had already suggested, certain smart predators—birds or lizards, for instance—would learn to avoid a noxious butterfly species based on its warning colors and behavior. But they'd still enjoy feasting on any unrelated, edible butterflies in the vicinity. Meanwhile, within each edible species genetic mutations that affected the butterflies' appearance would constantly arise and get winnowed out. This winnowing was accomplished by the predators: based on their experience with the noxious, warningly colored butterfly, these smart predators would decide which prey to attack and which to avoid. At some point, simply by chance, a new mutation arises within an edible species that causes some individuals to resemble the noxious model a bit more. This chance resemblance helps a few of these new mutants escape predation. Honed by smart predators, over time the edible species will accumulate mutations that steadily improve its fidelity to the noxious model. Eventually, the edible mimics look enough like the noxious model that predators will consistently avoid them. Like parasites, these mimics now reap the benefit of lower predation risk without having to invest in their own chemical weaponry. As the still-tasty mimics become more and more abundant, they begin to undermine the message originally conveyed by the warning signal. Modern biologists believe that Batesian mimicry systems are caught up in an evolutionary game of cat and mouse: as the noxious models escape their mimics by evolving new warning displays, the edible mimics continue to chase them.

But other mimics can be beneficial. Unlike Bates, Fritz Müller had no intention of ever returning when he left Germany in 1852. With his family, Müller immigrated to Brazil, where he farmed, taught at local schools, and somehow found time to closely observe the life around him. He noticed that co-occurring but unrelated noxious butterflies often seemed to converge on a common badge of distastefulness. Müller suggested that sharing a common warning display allowed several noxious species to cooperate in training their predators not to eat them. His reasoning was simple; Müller envisioned that a naive predator might need to eat one hundred warningly colored, noxious prey before it learns to recognize and avoid them. But if the predator can learn the same lesson by sampling from two similar-looking noxious species, it lowers the chance that any particular individual of either species will get eaten. In what is now known as Müllerian mimicry, the members of both mimic species enjoy a reduced per capita predation rate. Unlike Batesian mimicry, this mutualistic interaction spreads the communal pain over the whole well-defended species assemblage.

What about the scores of imitation fireflies? Some counterfeits, like the net-winged beetles, are most likely protected by their own brand of chemical defense. If so, we'd be looking at Müllerian mimics: by converging on common warning coloration, these noxious insects would help fireflies by splitting the cost of educating potential predators. Other mimics, Batesian style, might be perfectly tasty insects masquerading as fireflies to avoid getting eaten. And this particular evolutionary subplot gets even subtler. Many firefly species are also striking look-alikes: they've got a distinctive red and black pattern on their head shield, with pale lines edging their dark wing covers. To some degree, this resemblance flows from their shared ancestry. But could certain fireflies actually be tasty hitchhikers that evolved warning colors to mimic other, more noxious, fireflies? Or are *all* fireflies poisonous? We don't yet know.

## The Vampire Firefly

Like all mimicry systems, fireflies are deeply entangled in an evolutionary plot that we're just beginning to understand. Yet a splendid bit of scientific sleuthing led Tom Eisner and his team to uncover a truly extraordinary firefly mimic, one that might exceed your worst nightmare.

I've spent countless nights eavesdropping on intimate firefly courtships. My students and I go out early into the field to locate *Photinus* females, as these are hard to find. We're soon surrounded by hundreds of flying males, all silently sparking out their flashes. They, too, are searching for females. But *Photinus* females are coy. Perched down in the grass, they wait until they spot an especially attractive male before answering him with just a single, leisurely flash. Females deliver this response at a very precise time lag after the male's call, as described in chapter 3. Among different *Photinus* species, this so-called female response delay varies widely. Females of certain species respond after a half-second delay, while other females hesitate about five seconds. This female time lag is how a *Photinus* male recognizes his own females.

Over the years I've become pretty good at recognizing females' response flashes—usually I locate a female after she's given just one or two flashes. Yet all too often I've been completely hoodwinked. Peering down through grass blades, I find it's not an amorous female *Photinus* who's given that sexy response. Instead, I've just been duped by the deceitful flash of a firefly femme fatale, a mimic with purely carnivorous intentions.

In North America, certain fireflies in the group *Photuris* follow an unusual lifestyle: after they've mated, the females turn to catching, and eating, other fireflies. These insects employ surprisingly sophisticated hunting tactics, including a behavior known as aggressive mimicry. Because female *Photuris* mostly eat males, Jim Lloyd dubbed them femmes fatales. Fashioning their bioluminescent talents

**FIGURE 7.4** Sex or death? *Left*: Flashy and alluring, a predatory *Photuris* femme fatale attacks a hapless *Photinus* male (photo by Jim Lloyd). *Right*: Meal leftovers—just some chewed-up bits and a few legs.

into a deceptive lure, these clever females target the males of other firefly species by mimicking their own females' flashes. Once the unsuspecting male gets close, the femme fatale reaches out and grabs him. These insidious predators are large, long-legged, and agile, and they're capable of devouring several males each night.

One summer I spent some time watching these predators, up close and personal, as each one carried out her attack. Seizing the unfortunate male, the female wraps her legs around him in a tight and deadly embrace. Swiftly she sinks her strong jaws into his shoulder, and his blood—firefly blood is white—oozes from the wound. After exsanguinating him, she starts methodically chewing up his body parts, starting with the softer bits like his head, then moving on to his abdomen. Taking her time, she carefully masticates each mouthful, spitting out the hard bits as she goes. Within a few hours, only a few scattered bits remain from her meal. Jim Lloyd had once said to me, "If *Photuris* females were the size of house cats, most people would be afraid to go out at night." Now I knew exactly what he meant.

Lloyd showed that these femmes fatales are responsible for the deaths of many *Photinus* males. As he traveled around trying to decipher the flash codes of different *Photinus* species, Lloyd frequently found these fearsome creatures lurking in the grass. At each spot, he saw them very accurately mimicking the sexy response flashes given by females of whatever prey species was locally active. And they're versatile hunters. Choosing from her extensive repertoire of deceitful flashes, each predatory female can switch signals to match her prey. This aggressive lifestyle contrasts starkly with fireflies' reputation as gentle and ethereal creatures—food for thought next time you're out admiring the apparent tranquility of a night filled with silent sparks.

But what's going on here? Most fireflies abandon eating once they've become adults. What could possibly have driven these voracious femmes fatales to such carnivorous extremes?

This question lay dormant until Tom Eisner decided to widen his study of fireflies' chemical defenses. Testing a handful of firefly species, he and his team had discovered lucibufagins in three *Photinus* species, and also in a diurnal firefly, *Lucidota atra*. But after testing some *Photuris* fireflies they'd collected from fields around Ithaca, they noticed wide variation in how much toxin these particular fireflies contained. Some *Photuris* contained no lucibufagins at all, while others—all of them females—contained substantially more. Were *Photuris* femmes fatales getting more than just a good meal by feeding on *Photinus* males? Could they also be drinking in their prey's lucibufagins?

To answer this question the researchers would need some pristine *Photuris* females, ones that had not yet eaten any prey. So they brought field-collected *Photuris* larvae into the lab and raised them into adults. Then these culinary virgins were split into two groups. Some *Photuris* females were each given two *Photinus* males, which they quickly attacked and ate. The rest ate nothing. The researchers then measured each female's lucibufagin content. Taking advantage of fireflies' tendency to reflex bleed, they could gently pinch each female and analyze the small drop of blood she released. Sure enough, when *Photuris* females had consumed two *Photinus* prey, they had lots of lucibufagin in their blood, while the other females were essentially lucibufagin-free. These stealthy femmes fatales were not only deceiving and eating *Photinus* males, but they were also usurping their toxins. Why?

Eisner's team enlisted some generalist insectivores to help determine if *Photuris* predators might be stealing lucibufagins to ward off their own enemies. They used *Phidippus* jumping spiders, which, like hermit thrushes, were known to find *Photinus* fireflies repulsive. The spiders were offered field-collected *Photuris* females; again, half of the females had eaten two *Photinus* while the others had eaten nothing. Spiders didn't attack any of the *Photinus*-fed females, yet they attacked and ate about half the others. The femmes fatales were using their prey's toxins for their own self-defense.

As we've seen, *Photinus* fireflies have evolved a multifaceted defense strategy that safeguards them against many predators, from birds and bats to jumping spiders. So it's a rather ironic twist of evolutionary fate that these same protective defenses have turned them into a highly attractive target for this particularly treacherous mimic.

Unable to mount an effective chemical defense of her own, the *Photuris* femme fatale is compelled to spend her nights in pursuit of firefly prey to slake her thirst for toxins. But their aggressive flash mimicry is just one of several active hunting strategies that *Photuris* females use to obtain their lucibufagin fix. They also attack *Photinus* on the wing, targeting males by pinpointing their flash signals. For a *Photinus* male, this creates a tricky balancing act. Their own females prefer those males who emit more conspicuous courtship flashes, as described in chapter 3. Yet when a *Photinus* male flashes longer or more often, he's suddenly more vulnerable to predatory *Photuris*. Pulled in opposite directions by the two distinct evolutionary forces of sexual selection and natural selection, the courtship signals of these *Photinus* males are poised in exquisite balance.

And there's one last trick that *Photuris* females sometimes use to procure the lucibufagins they so desire: they simply resort to thievery. Stationing herself near a spider's web, a female waits quietly until some unlucky male firefly blunders into the web while he's searching for a mate. Soon the spider has neatly wrapped the male in silk, storing this tasty prey for later consumption. Now the *Photuris* female jumps lightly onto the web. She agilely steps over and grabs this gift-wrapped prize. By taking advantage of the spider's web-building talents, it seems this *Photuris* female has stolen herself an easy meal. Hunching over the immobilized prey, she begins to feed. Yet her audacious behavior isn't risk free, because the spider often returns to do battle. If the spider has a size advantage, this contest will end with the *Photuris* burglar also getting wrapped and eaten.

* * *

Their spectacular sex lives is what first got me hooked on fireflies. Lately, though, I've been fascinated by these stories of chemical defense—such a thrilling mix of poison, treachery, and thieves! To ward off hungry predators, fireflies have evolved a highly successful, multifaceted defense strategy. Through some biochemical wizardry we don't yet understand, fireflies manufacture potent toxins to poison their enemies. Recent discoveries have helped solve one long-standing mystery:

what function did firefly bioluminescence first serve? We've seen that fireflies' light provides a conspicuous and highly memorable advertisement that warns nocturnal predators to stay away from this toxic prey. Beyond bioluminescence, fireflies use other tactics to avoid attack: bright warning colors, pop-up glands that secrete noxious chemicals, and a repellent habit of oozing blood when they're injured. Driven by the exigency of survival in an insect-eating world, fireflies' fully featured defense package vividly illustrates the power of natural selection.

In a surprising plot twist, some fireflies seem to have lost the evolutionary knack for making toxins. Vulnerable to predators, *Photuris* fireflies must hijack other fireflies' chemical weapons and stockpile them to use in self-defense. Turning carnivorous, they rely on deceit and theft to obtain their toxins. These predatory fireflies have certainly flipped the proverb heading this chapter: everyone else's poison has become their meat.

Tom Eisner died in 2011, following a prolonged battle with Parkinson's disease. Insects were his passion, and he'd spent a lifetime devoted to revealing their secret chemical weapons. During an interview in 2000, Eisner joyfully recalled how he'd unveiled the story of firefly defenses—it was, he said, "a mystery in the night, and boy was it fun to work on!"

Clearly, many mysteries of firefly chemical defenses remain to be solved. Like those Russian matryoshka dolls where the outer doll must be popped open before the next figure reveals itself, nature's secrets are only slowly revealed. How do fireflies make—or take—such powerful toxins without poisoning themselves? Do males' nuptial gifts carry these valuable toxins? What stockpiles of chemical weapons have other fireflies amassed? Although he was hardly unbiased, Tom Eisner considered insects the most versatile chemists on Earth. Yet we've examined chemical defenses for only a tiny fraction of the 2,000 firefly species worldwide—fewer than 0.5%! Nature's creative chemists have fashioned a wealth of products that have become indispensible for human health—antibiotics, heart drugs, painkillers, anticancer agents. Who knows what chemical riches lie waiting to be discovered within fireflies' pharmacopeia? But the opportunity for such discoveries may be quickly running out.

CHAPTER 8

* * *

# Lights Out for Fireflies?

*Don't it always seem to go*
*That you don't know what you've got*
*Till it's gone.*
*They paved paradise*
*And put up a parking lot.*
- Joni Mitchell -

## Darkening Summers

As the wave of humanity washes over our planet, entire ecosystems have vanished. So it might seem rather inconsequential when one species or two go extinct. Each loss represents but a tiny pinprick in the fabric of life. Yet some species leave an unfillable void. If fireflies were to disappear, Earth's natural magic—and the quality of our lives—would be perceptibly diminished. Of course, it wouldn't happen all at once. It would be more like extinguishing a roomful of candles, one by one. You might not notice when the first flame is extinguished, but in the end you'd be left sitting in darkness.

When I mention my work, the question people most often ask is: "Why are all the fireflies disappearing?" To be sure, fireflies have good years and bad years, depending on local conditions. Yet most people are convinced there are fewer fireflies now compared to when they were growing up. A firefly watcher from Mulberry, Florida, e-mailed that she "grew up with such an abundance of them, but they have been gone for years now." Another firefly fan near Houston, Texas, wrote that she used to see "fireflies all over the place when I was a kid, but sadly they disappeared." And a Florida rancher observed that "they used to be everywhere. Now if you see three or four, you've really seen something." Firefly expert Jim Lloyd has also seen Florida firefly populations declining over the past few decades. And he's seen the same trend elsewhere as well. Wherever he travels in the United States, he's noticed "there aren't anywhere near as many as there used to be." People around the world are voicing similar concerns. In 2008, a Thai boatman who grew up amid the synchronous fireflies along the tidal rivers south of Bangkok estimated that "firefly populations have dropped 70% in the past three years." "I feel like our way of life is being destroyed," he lamented.

Skeptics might point out that appearances can be deceiving. Couldn't this perception be due to people's changing lifestyles instead of declining firefly numbers? Some of us who grew up chasing fireflies in suburbs or in the country have become city dwellers. And the advent of air conditioning means that fewer people spend summer evenings sitting out on their porch, enjoying the breeze and a cold drink. As technology marches onward, computers, video games, and mobile phones have captivated our collective eyeballs—we're now more likely to gaze into such ubiquitous screens than to gaze over a meadow or peer into a forest at night.

Yet a precipitous decline in firefly numbers over the past century has been quite thoroughly documented in Japan. And although we lack long-term population data for most countries, many serious naturalists and other attentive folks are also convinced that fireflies are on the decline. While the evidence is anecdotal, such a consistent refrain suggests that fireflies are vanishing from many places around the world. What's going on here? Although we don't know for certain, several culprits rise to the top of an ignominious list of things that might be responsible for declining firefly populations: habitat loss, light pollution, and commercial harvesting.

## Paved Paradise

In 2010, international firefly experts convened to craft a document known as the *Selangor Declaration on the Conservation of Fireflies*, which lists habitat protection as the top priority for preserving firefly populations. What makes for good firefly habitat? Fireflies like to live in undisturbed grassy areas, forests, marshes, and along stream banks. Fireflies enjoy a complicated life cycle, and most require moisture during each life stage; their eggs, larvae, and adults can quickly die from desiccation. Females seek out damp places to lay their eggs, which take a few weeks to hatch. After hatching, most fireflies spend between several months and two years in their earthbound larval stage. During this time their grub-like larvae live underground, crawling through soil in search of food. Clearly these larvae are unlikely to travel very far afield. The immobile pupal stage, during which fireflies transform themselves into adults, is also a subterranean affair. The upshot is that a firefly adult might emerge above the ground only a few meters away from where it first got deposited as an egg.

Fireflies also don't disperse very far during the few weeks of their adult lives. In contrast to other insects—dragonflies, for example—most fireflies are feeble flyers. While firefly males enthusiastically take wing each evening to search for females, they stay pretty local during their courtship excursions. And female fireflies generally don't fly much at all; in glow-worm species, a wingless female might travel only a few meters during her entire adult life.

Here's the good news: these sedentary habits heighten the chance that an established population will remain in the same place year after year, as long as the conditions are right. The bad news: when the going gets rough for fireflies, they can't just pack up and leave. Once their habitat is destroyed, the fireflies are gone too. If a breeding population is disturbed, the fireflies are unlikely to relocate somewhere else. When this one candle flame gets snuffed out, we say the population has gone locally extinct.

So the prime suspect in the disappearance of fireflies is the steady decline in the natural areas—the meadows, woodlands, and marshes—where they live. In the United States, firefly habitat has often been swallowed up by an advancing wave of residential and commercial development. Suburban sprawl has replaced firefly meadows and forests with houses, parking lots, and shopping malls. We don't need land-use maps to know this is happening—we see it all around us. Considering the poor dispersal ability of most fireflies, it's easy to understand how such habitat loss can diminish their numbers. In response to complaints from some residents of Houston, Texas, about the dearth of fireflies there, firefly expert Jim Lloyd characteristically grumbled, "you had them in Houston before they built the city on top of them and paved it over." Lloyd also blames this loss of natural habitats for the disappearance of many firefly species around Gainesville, Florida, based on his nearly fifty years worth of experience. He watched as more than a dozen firefly species that were abundant when he first arrived there in 1966 vanished completely by the late 1990s. Exponential residential and commercial growth has wiped out most of the wetlands that make for prime firefly habitat. Increased water demand and farming practices have lowered water tables and dried up many of the marshes, creeks, and wetlands fireflies favor.

And fireflies elsewhere are equally vulnerable. One of my favorite species is *Photinus marginellus*, a tiny but appealing firefly whose early evening activity and knee-high courtship make it quite accessible. Over many years, we've studied a population outside Boston that's restricted to a small grove of cherry trees. Remarkably, their entire life cycle from egg to adult and back again to egg appears to be carried out beneath these trees. Luckily for them their habitat sits on conservation land that's protected from development, so they will continue to thrive.

But when soil gets bulldozed, removed, and replaced during construction and landscaping, even places that *look* like great habitat can be devoid of fireflies. A few years back I attended a family wedding at a spanking new golf resort in Delaware. In addition to participating in the celebration, I was excited to journey south into the territory of *Photinus pyralis*, the Big Dipper firefly. Optimistically, I brought along my headlamp and insect net. We arrived to find the golf course surrounded by acres of beautiful grassland, a habitat that looked ideal for these lawn-loving fireflies. So I had high expectations when I ducked out from the reception with some young cousins at dusk. Yet we found not a single firefly that night. Two years of construction activity had imported, relocated, and disturbed the soil to such an extent that no fireflies had survived. Such massive landscaping activity will not only affect firefly larvae directly but can also kill off their prey—the earthworms, snails, and other insects that young fireflies feed on. Pesticide use was also a suspect at this site, as manicured landscapes often depend on heavy use of such chemicals. If the firefly deficiency at this Delaware golf resort resulted only from construction activity, we can hope that some fireflies will eventually colonize and become established in this new habitat.

The loss of suitable firefly habitat is problematic not only in the United States but also worldwide. In Thailand and Malaysia, congregating *Pteroptyx* fireflies are recognized national treasures, and both of these Southeast Asian countries have developed thriving tourism industries based on these fireflies' nocturnal courtship displays.

At night, the tiny riverside village of Kampung Kuantan in Kuala Selangor is transformed into a major tourist destination. Here along the Selangor River near the west coast of Peninsular Malaysia, visitors climb aboard local sampan boats to glide silently through tidal channels. They've come to watch displays of the synchronous firefly *Pteroptyx tener*. As the sun sets, male fireflies take up perches atop mangrove leaves. Flashing randomly at first, then gradually becoming synchronous, the males' lights are reflected in the dark waters. Up until the 1970s, the Kampung Kuantan firefly populations were known only to local villagers and a few curious scientists. Now more than 50,000 tourists annually come to admire to this year-round Christmas tree display. Firefly ecotourism makes an important contribution to this village's economy, which otherwise relies on small-scale agriculture and fishing.

The best firefly habitat stretches for ~10 km along the Selangor River estuary, surrounded by flat coastal plain. Although many different kinds of mangrove grow along these tidal rivers, fireflies favor the berembang, *Sonneratia caseolaris*, as sites for their courtship and mating displays. After mating, *Pteroptyx tener* females fly away from the display trees and search along the riverbank for moist soil to lay their eggs. The eggs hatch within three weeks into crawling larvae, which spend a few months feeding on mangrove snails that thrive in damp leaf litter. When it has finished feasting, each larva digs a hole in the mud to pupate. Completing the cycle, the adult firefly crawls out from its pupal casing and flies off to join the congregation in a display tree along the river.

Once, the Selangor River was lined with mangrove trees. Now, large swaths of this native forest have been cut down and replaced with oil palm plantations (Figure 8.1). Palm oil has turned into a lucrative commodity on the global market, and Malaysia has become one of the top producers of this popular vegetable oil. Another threat to the mangrove forest is shrimp aquaculture farms, which are constructed by clearing large stretches of riverbank.

As these activities spread, habitat suitable for fireflies shrinks. By removing the mangroves, development puts the survival of Malaysia's congregating fireflies at risk during two distinct life stages. First, the larval fireflies: their habitat is being destroyed, and the snails they eat are being wiped out. The second part of this one-two punch hits the adults: the display trees used as congregation sites for courtship and mating are being destroyed. Surveys conducted along a 9-km stretch of the Rembau-Linggi estuary in 2008 and again in 2010 showed that destruction of mangroves along the riverbank had reduced the number of display

**FIGURE 8.1** Destruction of firefly habitat along the Selangor River in Malaysia, where riverside vegetation is being cleared to establish oil palm and banana plantations (photo by Laurence Kirton).

trees for synchronous fireflies from 122 to fifty-seven, a precipitous drop over merely two years.

While admiring nature is a commendable pastime, tourism itself can negatively impact firefly populations. A rapid explosion of firefly tourism in Malaysia and Thailand has led to unregulated commercial development and overexploitation along some firefly rivers. New resorts and restaurants catering to tourists have sprung up, pushing aside former firefly habitat. These establishments often use bright outdoor lighting at night, and this can interfere with firefly mating rituals. At one tourist destination in the Thai province of Samut Songkram, the number of firefly-watching boats skyrocketed from seven to 180 in just six years. As tourism boomed, diesel-powered tour boats contributed to water pollution and riverbank erosion. Some residents along the riverbank were reportedly so disturbed by nightly noise from all the tour boats, they chopped down the firefly display trees near their houses. Also problematic is the fact that some tour guides and boat drivers don't know much about the life cycle or habitat requirements of the creatures on exhibit. To entertain tourists, drivers sometimes crash their boats into trees, dislodging the fireflies into the water. They shine spotlights into display trees or capture fireflies to show off the insects. Obviously, such antics disrupt the fireflies' courtship and mating activities.

Sonny Wong, a senior conservation officer with the Malaysian Nature Society, is a lanky, athletic man with an easy smile and self-effacing manner. He's also an expert on Malaysian fireflies. Since 2003, Wong has worked to increase public awareness of firefly ecology and conservation needs. He's also been spearheading efforts to get local communities involved as stakeholders in firefly conservation efforts that will protect sensitive firefly habitat. The Malaysian Nature Society hopes to develop a sustainable firefly tourism industry by developing best practices and disseminating these guidelines among the local villagers. As Wong points out, these programs will "teach them the ethics of firefly-watching, and . . . provide them with a means of livelihood. They can also feel a sense of ownership over the area." In Thailand's Samut Songkram province and elsewhere, environmental educators now post signboards and distribute pamphlets answering firefly FAQ to local residents, tour guides, and boat operators. Moving forward, such community-based tourism will be key to balancing economic development against environmental degradation. To ensure that future generations will be able to experience what is surely a wonder of the natural world, let's hope that many stakeholders will embrace stewardship of these congregating fireflies.

## Drowning in Light

The tiny Swiss village of Biberstein is home to a vigorous population of *Lampyris noctiluca*, the common European glow-worm. Like other villages, towns, and cities around the world, it also has streetlamps. Stefan Ineichen, an educator and firefly admirer, decided to see how these artificial lights affected Biberstein's glow-worms. When they mapped out the display locations of glow-worm females, which are flightless, they found that the bright circles created by streetlamps did not affect where these females chose to perform their male-attracting glow show. Yet the streetlamps did change where males chose to conduct their search flights. The nonluminous, flying males only looked for females around the darker places away from bright lights. As a result, all the lonely female glow-worms who displayed their wares near streetlamps probably never got a chance to mate. So it seems the streetlamps of Biberstein and other European cities might unintentionally be punching out holes—like those in Swiss cheese—from the glow-worm's reproductive landscape.

**FIGURE 8.2** Light pollution adversely affects fireflies because it interferes with the bioluminescent signals they use to find their mates (photo courtesy of NASA).

For the last two hundred years, sheer human ingenuity has enabled us to conquer darkness. At night, artificial electric lighting helps us illuminate city streets and roadways, hold outdoor sporting events, advertise our wares, improve security in parking lots and around buildings, and dramatically highlight the specimen trees in our yards. Satellite photos reveal tentacles of streetlights that accompany roadways as they snake out into the wilderness.

To be sure, artificial light at night is usually beneficial. But light pollution creates problems. Man-made sources often produce stray light that shines where it's not intended. The International Dark-Sky Association estimates that 30% of all outdoor lighting in the United States is directed skyward, where it does no good. Poorly designed lighting is pervasive. It encroaches on darkness and changes our planet's natural light cycle. In the 1960s, astronomers first raised the alarm about light pollution, dismayed at how it impairs our ability to observe the splendors of the night sky.

And ecologists, too, have reason for concern over light pollution. Artificial lighting disrupts the natural behaviors of all sorts of nocturnal creatures, including birds, turtles, frogs, and insects. Light pollution has potentially devastating consequences for fireflies too, because their bioluminescent mating signals are easily drowned out by artificial illumination. Lights that shine into firefly habitat will increase the background noise for detecting courtship signals of both sexes. This lower signal-to-noise ratio may explain why, as the Swiss glow-worm study showed, firefly males tend to avoid searching for mates in brightly lit areas. Similarly, in a Florida firefly, *Photinus collustrans*, female decoys that were placed near bright outdoor lights were less successful at attracting males than those in darker spots. Artificial lights might even prevent courtship, since male fireflies rely on natural light cues at dusk to start up their search flights. In laboratory experiments, artificial light disrupted the courtship of Thai fireflies and lowered their mating success.

By jeopardizing their reproductive success, light pollution puts firefly populations at risk. For more than twenty years, Tennessee naturalist and firefly aficionado Lynn Faust has been making careful observations on a dozen different firefly species on her forty-acre farm outside of Knoxville. After a new house went up next door—a McMansion with thirty-two outdoor floodlights—she noticed that one resident firefly species had vanished. And it hasn't returned. "Why would you need so many lights?" ponders Faust. Since then she's organized a neighborhood campaign to reduce light trespass, when unwanted light from a neighbor shines into someone else's yard or home.

In a relatively short time, human technology has completely transformed the nighttime: natural darkness has now vanished from much of the Earth's surface. This stray light blinds us to the beauty of the night, including fireflies. I find myself so addicted to artificial light that it takes some determination for me to walk down a forest path at night without a flashlight. But when I do, so many glowing things are revealed to my dark-adapted eyes! Suddenly I can see firefly larvae dragging their slow glows across the ground, sometimes flocking together in hundreds. Here's a firefly pupa nestled cozily into the ground, its entire body shining. And here's a toad hopping along and flashing—has it swallowed a firefly?

As we light the world to suit our needs, we should keep in mind that this illumination comes at the expense of other living beings. If we want more fireflies around us, we need to take back the night. Some ideas about how to accomplish this by making smarter lighting choices are provided at the end of this chapter.

## A Bounty on Firefly Lights

Bounty hunting could also be a reason for vanishing firefly populations. For nearly half a century, fireflies in the United States were harvested in vast numbers from wild populations to extract their light-producing enzyme, luciferase. While luciferase helps fireflies talk to one another, this enzyme has proven useful to people for entirely different purposes.

Luciferase is found in firefly lanterns. As described in chapter 6, this enzyme brokers a chemical reaction that gives off light by harnessing energy contained in a molecule called adenosine triphosphate, or ATP. All living things—from bacteria and slime molds to fireflies and humans—have ATP, which they employ as a chemical courier for transporting energy around inside their cells. As long as something is alive, you'll find ATP inside it. The discovery that firefly light is powered by ATP was made in the late 1940s by William McElroy, a biochemist working at Johns Hopkins University. The fact that firefly luciferase only glows when ATP is around meant that luciferase could be used to tell whether particular cells were dead or alive. McElroy's discovery led directly to many practical uses for luciferase in medical research and food safety testing. Since living fireflies were once the sole source of luciferase, firefly hunting became a popular summer pastime, innocently pursued by a small army of children who were paid to go outdoors and harvest fireflies. Slaughtering fireflies to detect life carries a certain irony, but this is a true story.

Large-scale firefly harvesting began back in 1947, when McElroy's lab in Baltimore was trying to solve the mystery behind firefly bioluminescence. Their experiments needed firefly luciferase, which the lab extracted by grinding up the lanterns of lots and lots of fireflies. At first, the scientists themselves collected whatever fireflies they needed. But soon their experimental needs exceeded their own collecting ability. So they placed advertisements in the local newspaper to enlist the help of local children, who got paid 25 cents for every hundred fireflies they delivered. In the first year, they took in some 40,000 fireflies. By the 1960s, McElroy's lab was paying children to harvest between 500,000 and a million live fireflies each year. As a local newspaper article later put it, "Life in Baltimore, for fireflies, became a chancy, precious thing." McElroy knew enough about firefly natural history to realize that his child brigade would mainly be capturing the male fireflies. For the most part, the females would remain safely on the ground. As long as these females could still find mates and lay their eggs, he reasoned, this science-sponsored firefly harvesting might not have too much of an impact on firefly populations. In and around Baltimore, whatever fireflies were being collected were likely to be the Big Dipper *Photinus pyralis*, an easily identifiable species that was—and still is—abundant in this area.

But McElroy's firefly-harvesting operation turned out to be trivial compared with what lay ahead. Stemming from the discovery that luciferase glowed in the presence of ATP, new practical applications for firefly luciferase were rapidly being developed. Soon the Sigma Chemical Company of Saint Louis, Missouri, began selling luciferase, which they obtained from live fireflies by freeze-drying them and cutting off their lanterns. In the summer of 1960, the company rolled out what they called the Sigma Firefly Scientists Club, which would eventually enlist a vast network of firefly collectors to harvest many millions of wild fireflies nationwide. Each summer, Sigma flooded newspapers across the United States with advertisements announcing that fireflies were urgently needed for medical research (Figure 8.3). The Firefly Scientists Club, they said, was "open to all, Boy Scout groups, church groups, 4-H clubs, and individuals." If you sent in live fireflies, Sigma offered a bounty that began at 50 cents for the first hundred fireflies, increasing to a penny each for quantities greater than 20,000 fireflies. And you could earn up to a $20 bonus for capturing more than 200,000 fireflies.

Many well-intentioned families, kids, and community groups might have gotten hooked by Sigma's advertisement explaining that fireflies would be used to "diagnose human ailments, look for life on other planets, and fight pollution of

CASH FOR FIREFLIES

One Dollar Per Hundred

"Lightening Bugs" are urgently needed for Medical Research. Start a group or collect as an individual.

For Further Information

Call or Write

Sigma Firefly Scientist Club

3500 DeKalb Street
St. Louis, Missouri
63118
771-5765

**FIGURE 8.3** Newspaper ad for firefly bounty hunting by Sigma Chemical Company from the June 11, 1979, *Southeastern Missourian*.

air, food, and water in our own environment." Over the years, the Firefly Scientists Club blossomed into an enormous firefly-harvesting network with thousands of collectors distributed across twenty-five states, mainly in the Midwest and eastern United States.

Collectors from Illinois and Iowa usually led the pack of firefly hunters. One Iowa woman, nicknamed the "Lightningbug Lady," caught and sold to Sigma around a million fireflies every year for decades. Some of these fireflies she captured herself, trawling with her net while driving around in a pickup truck, but mainly she packaged and shipped off fireflies gathered by her local network of 420 firefly catchers. And one collector traded in fireflies for a different kind of summertime fun: she gave up her firefly earnings to help build a community swimming pool.

And so, believing they could "net extra cash and serve science at the same time," members of the Firefly Scientists Club captured and shipped off to Sigma an awful lot of live fireflies. How many fireflies? During the 1976 season, Sigma reported receiving 3.7 million fireflies, and for 1980 this number was 3.2 million. Although there's no way of knowing the exact toll taken by this commercial harvesting, a conservative estimate based on annual collections of three million fireflies for about thirty years yields a total of ninety million fireflies. That's a lot of fireflies.

So what happened to all those fireflies? Sigma (which would later become Sigma-Aldrich Chemical Company) processed these beetles and then sold a variety of luciferase products to be used for ATP assays. Research scientists, government agencies, and testing laboratories bought these products. The company's website and catalog still list a panoply of products from these harvested fireflies, including (with 2013 prices): whole desiccated fireflies (5 grams for $79), desiccated firefly lanterns (5 grams for $1,245), firefly lantern extract (50 milligrams for $183), and purified firefly luciferase (1 milligram for $186). There's no doubt that the Firefly Scientists Club netted big profits for Sigma-Aldrich. This company has gone on to become one of the world's largest biochemical suppliers, with international sales exceeding $2 billion in 2007.

In the mid 1990s, the Firefly Scientists Club said they were no longer accepting new members, and the company has apparently shut down its collecting operation. I hope this was partly due to the negative publicity that followed an interview I gave in 1993 to the *Wall Street Journal*, where I pointed out the harmful effects that Sigma's large-scale harvesting could have on firefly populations. I also pointed out that harvesting wild fireflies was no longer necessary due to a nifty technological advance. In 1978 scientists figured out how to isolate and determine the DNA sequence of the gene responsible for producing luciferase in *Photinus pyralis* fireflies. Once this genetic blueprint for luciferase had been discovered, scientists could manufacture luciferase without harming a single firefly. Using recombinant DNA techniques, the luciferase gene could be inserted into harmless bacteria whose protein-assembling machinery then cranked out large quantities of luciferase. This synthetic luciferase has been available since 1985, and it's cheaper to produce and more reliable than what's extracted from live fireflies. As a result, there's no longer any reason to harvest fireflies from wild populations. However, as recently as summer 2014, a mysterious little business called The Firefly Project based in Oak Ridge, Tennessee, was still actively harvesting fireflies from the wild. Again, they put ads in local papers soliciting collectors, who harvested about 40,000 fireflies from a single Tennessee county and got paid a total of $665.

I'm an ecologist, so it seems obvious to me that such massive firefly harvesting could not be sustained without wiping out some local firefly populations. Yet I'd guess that Sigma's collectors truly believed that fireflies were an inexhaustible natural resource—just like passenger pigeons, which are now extinct. What made matters worse was that the collectors had neither the ability nor the incentive to distinguish one firefly species from another—all fireflies earned them equal cash. Sigma marketed their fireflies and their luciferase as being from "*Photinus pyralis*"—after all, that's what McElroy's brigade had collected around Baltimore. Yet firefly collectors nabbed anything that flashed in the nighttime, so quite a few species were undoubtedly caught in their nets. Sigma also did not distinguish between common and rare species, although a representative noticed that some of the fireflies they received were "big and active" (these were probably *Photuris*), while others were "quiet" (these likely included several different species of *Photinus* as well as *Pyractomena*). In the next section we'll learn some easy ways to distinguish among these different groups of fireflies, but Sigma indiscriminately processed whatever fireflies they received.

How did bounty hunting affect US firefly populations? As described earlier in this chapter, fireflies are not good at dispersing to new habitats, so each local population stays local. By removing thousands of fireflies from any one habitat, collectors certainly depleted male firefly numbers and reduced females' mating prospects. Fewer eggs were laid, which meant fewer larvae hatched. Repeated collecting year after year in the same place would cause firefly numbers to steadily drop. While it's possible that a few very abundant firefly species might have tolerated such massive harvesting, many rarer species were probably exterminated. But what *are* sustainable levels of firefly harvesting? My colleagues and I used a computer model to answer this question by setting up virtual firefly populations and subjecting them to different degrees of harvesting. Although we needed to make some biological and mathematical assumptions, our results suggest that whenever annual harvest rates exceed 10% of the adult males, a typical population of *Photinus* fireflies would be driven to extinction within fifteen to fifty years.

Overharvesting has put fireflies at risk in other parts of the world, too. In the last section of this chapter, we'll see that commercial harvesting for aesthetic purposes during the nineteenth century resulted in some Japanese fireflies nearly being loved to death. And in China, a tourist park in Shandong Province imported 10,000 fireflies in 2013 to attract visitors. But within days, delight turned to dismay when half the transplants died because their new habitat was not well suited for fireflies. Perhaps other countries will learn from these mistakes.

## Other Insults

Fireflies also face other hazards that might contribute to declining populations, including pesticides. In many parts of the world, both soils and water are contaminated with high pesticide levels. In the United States, suburban lawns and gardens have up to threefold higher rates of pesticide application per acre than agricultural fields. Many of the insecticides commonly used on lawns are broad spectrum, meaning they're designed to kill whatever insects they contact. They don't distinguish between harmful ones, like Japanese beetles, and harmless ones, like fireflies. And remember that during their egg and larval stages, fireflies spend a lot of time underground, where they are likely to come into contact with insecticides. Adult fireflies also get exposed to insecticide residues when they rest on vegetation during the day.

Surprisingly few scientific studies have directly investigated how pesticides might affect fireflies. But a 2008 Korean study tested common insecticides to see if they harmed *Luciola lateralis* (now renamed *Aquatica lateralis*). This study found that nearly all the insecticides, when applied at concentrations recommended by the manufacturer, were highly toxic: many caused 100% mortality in the eggs, larvae, and adults of this firefly. Pesticides can also indirectly harm fireflies by killing off the earthworms and snails that firefly larvae eat. For example, the systemic herbicide 2,4-D, which is contained in products like Weed & Feed, has been shown to be toxic to earthworms, as well as to beetles such as ladybugs. Some Japanese scientists have suggested that widespread use of pesticides in Japanese rice fields contributed to the decline of fireflies there. So indiscriminate use of pesticides on lawns and gardens is likely to have adverse effects on fireflies.

We don't yet know how fireflies will be reacting to climate change. Rising temperatures will cause insects in the temperate zone to develop more rapidly and will also increase their overwintering survival. The seasonal activity period for many insects will last longer. Warmer temperatures and longer growing seasons might allow some insects to squeeze in more generations each year. Good news if the insects are fireflies; bad news if they're mosquitoes or other pests.

Fireflies, like many seasonal creatures, rely on temperature to cue their emergence times. Climate change has already caused shifts in many natural events that are based on thermal cues. Cherry trees in Japan are blooming earlier, birds are migrating sooner from their overwintering habitats, and frogs are breeding earlier. The same thing is happening to fireflies, it seems. For twenty years Lynn Faust has been carefully tracking the dates when the synchronous fireflies of the Great Smoky Mountains, *Photinus carolinus*, first appear and when they reach their peak density. These fireflies now put on their peak display about ten days earlier than they did twenty years ago.

Higher temperatures will also shift fireflies' geographic range toward higher latitudes. Yet the range of many species will probably shrink overall, because the southern portions of their range may become unsuitable. As rainfall patterns shift, fireflies will probably be unable to survive in areas with extended drought. As we come to terms with our future, there's little doubt that it will be a different world for fireflies, just like for the rest of us.

## HOTARU KOI: COME FIREFLY!

When I first visited Japan twenty years ago, I was startled by the deep affection the Japanese people show for insects, which are known as *mushi*. Everyone—from toddlers, to teenagers, to the elderly—is fascinated by insects. In stark contrast to Western entomophobia, the cultural norm in Japan is enthusiastic entomophilia. Japanese children eagerly embark on insect-collecting expeditions with their families, and even toddlers know how to accurately identify many insects. Live beetles are popular pets, and these can be purchased in upscale department stores and from vending machines.

Although beloved by people all over the world, fireflies, or *hotaru* (written 蛍 in kanji, ほたる in hiragana), hold a special place in Japanese culture. For a thousand years, they've been celebrated in art, poetry, and myth. So it was an especially poignant loss when Japanese fireflies were nearly extinguished in the twentieth century due to habitat degradation and overharvesting. But diligent efforts to restore habitats and reintroduce fireflies have transformed a predictably sad saga into a resounding success story for environmental conservation.

Among the fifty different fireflies in Japan, two are especially beloved. The larger one is known as the Genji firefly (*Luciola cruciata*), which lives near rivers and fast-flowing streams. The Heike firefly (*Aquatica lateralis*) is smaller, and lives near rice fields and other stagnant water. Both species are intimately connected to aquatic habitats because these fireflies spend their entire larval stage living underwater. Females lay their eggs in mossy areas adjacent to streams. The newly hatched larvae crawl down into the water, where they spend several months feeding exclusively on freshwater snails. It's not until these aquatic larvae get ready to pupate that they crawl back onto land again, where they pupate in moss-covered soil along the river's edge. The arrival of their bright lights is a hallmark of summer. When fireflies are abundant, they sometimes all flash in unison, their slow dancing lights floating silently above the water.

Once, fireflies were abundant throughout Japan. Especially for aquatic fireflies, Japan's many mountains, rivers, streams, marshlands, and irrigated rice fields provided nearly perfect conditions. During the Edo period (1603–1867) hotarugari 蛍狩 (ほたるがり), or firefly catching, was quite a fashionable summer activity. Many beautiful woodblock prints and paintings depict children and grownups alike chasing and catching fireflies with fans, nets, and bamboo cages. Noble families hosted elaborate firefly-catching parties and excursions as an aristocratic diversion. Yet even the most indigent farmer could enjoy this free activity. The popularity of firefly catching continued during the Meiji period (1868–1912), when children all over Japan sallied forth on moonless summer nights for firefly hunts. To attract their shining prey, children chanted songs while hunting: "Hotaru Koi (Come Firefly)" has many regional variations, but one translation goes:

**FIGURE 8.4** *Love of Fireflies*, a 1896 lithograph by Yosai Nobukazu (from the private collection of Ross Walker, www.ohmigallery.com), captures the deep connection between fireflies and the Japanese people.

> Firefly, come hither, and you shall have water to drink!
> Yonder the water is bitter; here the water is sweet!
> Come, fly this way, to the sweet side!

During the season, tourists would flock to places renowned for their spectacular firefly displays. The town of Uji is one of the most celebrated tea-growing regions of Japan, but in the early 1900s Uji was equally famous for its fireflies. Each summer, thousands of visitors arrived on special trains that ran from Kyoto and Osaka. When the light show reached its peak in June, *hotaru-bune* boats made nightly excursions along Uji's river, the Uijwara, where their passengers had picnics while viewing the fireflies. In 1902 Lafcadio Hearn, acclaimed author and interpreter of Japanese culture, described this summertime spectacle:

> The stream there winds between hills covered with vegetation; and myriads of fireflies dart from either bank, to meet and cling above the water. At moments they so swarm together as to form what appears to the eye like a luminous cloud, or like a great ball of sparks. The cloud soon scatters, or the ball drops and breaks upon the surface of the current, and the fallen fireflies drift glittering away.

Toward the end of the evening, he went on, "the Ujiwara, covered with still sparkling bodies of the drifting insects, is said to resemble the Milky Way." But these sparkling insects would soon evaporate like glowing smoke.

In the next few decades, firefly catching for diversion would segue into firefly hunting for profit. Despite a deep appreciation for the natural world, the Japanese were not hesitant to exploit their natural resources. Fireflies were in vogue, and live ones were worth good money. Firefly-collecting shops sprouted up in prime locations, each hiring dozens of firefly hunters. These men worked sunset to sunrise collecting live fireflies from May to September: a skilled hunter could bag 3,000 fireflies in a single night. Each morning the fireflies were carefully packaged into wooden cages with some damp grass. Then they were dispatched via express messengers to clients in Osaka, Kyoto, and Tokyo: hotel proprietors, restaurant owners, private citizens. Once the fireflies reached their destination, the brightly glowing insects were released into hotel gardens and restaurant courtyards for customers to admire their luminous displays.

These city dwellers undoubtedly made a highly appreciative audience. But massive numbers of Japanese fireflies were being extracted from their native habitats—these treasured insects were getting loved to death! Their demise might have been hastened by certain firefly-hunting tactics, which Hearn described:

> As soon as the trees begin to twinkle satisfactorily, [the firefly-hunter] gets his net ready, approaches the most luminous tree, and with his long pole strikes the branches. The fireflies, dislodged by the shock . . . drop helplessly to the ground . . . where their light—always more brilliant in moments of fear or pain—renders them conspicuous. . . . Thus the firefly catcher works until about two o'clock in the morning . . . at which time the insects begin to leave the trees and seek the dewy soil. There they are said to bury their tails so as to remain viewless. But now the hunter changes his tactics. Taking a bamboo broom he brushes the surface of the turf, lightly and quickly. Whenever touched or alarmed by the broom, the fireflies display their lanterns, and are immediately nipped and bagged. A little before dawn the hunters return to town.

Although firefly gender seemingly went unnoticed, the females of Genji fireflies are known to gather together after midnight to lay their eggs along the mossy riverbanks. Most likely, the fireflies that these hunters were collecting from two in the morning until dawn were females "burying their tails" as they deposited their eggs. By targeting these egg-bearing females, this hunting tactic was extinguishing the sole chance the firefly population had to replenish itself.

And so it was that by 1940 people began to notice that firefly populations around the country were blinking out. Several factors—not just commercial hunting—contributed to this widespread decline. One problem was the river pollution that accompanied rapid development of Japanese industries and urban centers. Water quality suffered as industrial effluent, agricultural runoff, and household sewage flowed into rivers. This river pollution reduced survival for aquatic firefly larvae and also for their snail prey. An additional problem came in the form of government-sponsored river canalization projects; installing concrete embankments for flood control destroyed the mossy riverbanks where female fireflies prefer to lay their eggs and where the larvae crawl out to pupate.

Yet the demand for live fireflies persisted, so the commercial firefly houses decided to try breeding fireflies in captivity. Watching carefully as fireflies passed through each life stage, these breeders used pure trial and error to determine how to raise Genji and Heike fireflies indoors. Luckily, fireflies that have aquatic larvae are easier to breed in captivity than those with terrestrial larvae. As part of

**FIGURE 8.5** With their long-lasting glows, Genji fireflies float above a river in Japan (*Luciola cruciata* photo by Tsuneaki Hiramatsu).

pinpointing the conditions most conducive for fireflies' survival, the breeders determined exactly what snails the carnivorous juveniles like to eat, and exactly what mosses the females prefer for their egg laying. By the mid-1950s, many indoor breeding facilities had been set up, and these provided fireflies both for sale and for reintroducing into streams and rivers. These breeding efforts also brought to light many previously unknown details concerning the ecology and life cycle of these Japanese fireflies. Scientists have since worked out methods for breeding several other Asian fireflies, although so far only those species that have aquatic juveniles have been successfully bred. Today, captive-bred fireflies can even be purchased online in Japan.

Dr. Norio Abe runs the Tokyo Firefly Breeding Institute, located in an inconspicuous low building in Itabashi Ward. Abe-san comes from a family that for many generations has specialized in brewing fine sake, or rice wine, but he now specializes in working out the perfect formula for brewing Genji fireflies. When I visited the institute one June, Abe-san greeted me at the entrance with a wide smile. After we exchanged greetings, he ushered me in to see his firefly magic. In a room cluttered with freshwater aquaria was everything that a firefly might need to succeed during each stage of its life. Hundreds of firefly larvae were happily wriggling around underwater in an oversized tank. Peering at them through the glass, I could see that many were busy feasting on their favorite snail prey, *Semisulcospira libertina*. Older firefly larvae were housed in another tank—here, earthen banks sloped down into the shallow, well-aerated water. These larvae were almost ready to pupate, and Abe-san described how they would soon crawl out to metamorphose in the moist soil. More tanks with grow lights held moss—the special kind that females like to lay their eggs in; others housed snails and their food. At the rear of the institute we entered the warm, moist air of a long greenhouse complete with trees, shrubs, and mist-making devices. An artificial brook babbled noisily down the center. Later that summer, the winged adults of Abe-san's well-bred fireflies would be set free into this enclosure to fly, to flash, and to mate. And for one night the institute would invite city dwellers to enjoy a special firefly viewing—an authentically Japanese experience, even if only within these glass-walled confines.

Beginning in the late 1970s, drastic declines in firefly populations inspired widespread habitat restoration efforts in cities around Japan. Galvanized into action by the loss of these summertime icons, many local communities undertook municipal projects to clean up their rivers and restore suitable habitat for larval fireflies and their snail prey. Sewage treatment plants were built, ordinances were enacted to control industrial and agricultural effluent, and riverbank areas were reengineered specifically to make them more firefly friendly. With many firefly rivers officially protected, cities such as Osaka and Yokosuka City set up firefly breeding programs to rear fireflies from eggs. Thousands of captive-bred larvae were released in efforts to repopulate rivers. These firefly restoration projects enjoy strong public support throughout Japan, with local schoolchildren, the elderly, and other volunteers enthusiastically participating in such efforts. The revival of Japanese fireflies has been remarkably successful and has helped to raise environmental awareness throughout Japan. Now, fireflies have become a national symbol for successful environmental conservation.

Many cities and towns throughout Japan hold annual firefly-watching festivals known as *hotaru-matsuri*. Although fireflies show up in greatly reduced numbers compared to the olden days—3,000 fireflies now make a respectably sized population—families, photographers, and young lovers still come to admire their glows. Popular destinations during June and early July, these festivals make a sizable contribution to the local economy in many towns. And these festivals celebrate not just the magic of these silent sparks but also the collective effort that made their reincarnation possible.

Some years back I was honored by an invitation to speak at the Santo Hotaru Firefly Festival hosted by the town of Maibara, Shiga Prefecture. This is one of the prime firefly-viewing spots in all of Japan, and the Santo firefly festival has been held there each June for decades. Here, the Amano River has been designated as a Special Natural Monument to protect the fireflies and their habitat. (A few other countries—including Malaysia and China—have also established preserves to protect firefly habitats.) During firefly season, volunteers from the local firefly conservation group carefully monitor the weather conditions and firefly numbers at several sites along the river. The same group also runs school programs to teach the next generation about firefly biology, life cycles, and conservation (Figure 8.6). This firefly festival attracts many visitors and includes a children's parade, food vendors, and many stalls selling firefly-related items. Car traffic is strictly regulated during the festival, and a shuttle bus service brings people to the firefly viewing area, where firefly catching is strictly forbidden.

My companions and I arrived after sunset to set up in the auditorium. The lobby displayed posters explaining the firefly life cycle and many details of their behavior, along with historical photographs and some charming drawings by local schoolchildren. Many townspeople—ranging in age from eight to eighty-eight—came to hear my talk on American fireflies, and afterward I answered questions from the audience. They showed an impressive depth of knowledge concerning Japanese fireflies and were fascinated to hear about firefly diversity in other parts of the world. Later that night we stepped out into a gentle rain to see the fireflies. Local volunteers wearing firefly-decorated safety vests guided us down to the river. Along the way, we noticed that all the streetlamps had been fitted with shields to prevent stray light from disturbing the fireflies. When we arrived, the Genji fireflies with their long-lasting glows were calmly floating above the river like green embers. And everyone wanted to gently hold a firefly before releasing it back into the sky.

**FIGURE 8.6** An educational drawing exhibited at the Maibara Firefly Festival succinctly explains what both human and firefly babies eat (*left*) and where they live (*right*).

Although fireflies disappeared from Tokyo long ago, the summer of 2012 ushered in some hi-tech substitutes. The Tokyo Hotaru Festival was "created with the desire of bringing fireflies to Tokyo and allowing residents to coexist with nature." During this event, a flotilla of 100,000 solar-powered, glowing ping-pong balls is released into the Sumida River as it flows through downtown. Although seemingly a poor substitute for the vanquished living sparks, in 2013 nearly 280,000 people attended this popular event.

There's a Japanese proverb—"Children grow up watching their parents' back." Each generation trods cultural pathways worn smooth by many feet. The Japanese believe that humans and nature are part of one whole, and both must adapt to a world that is rapidly changing. Some might view this flotilla of "fireflies" as a ghostly reminder of days gone by. Yet this futuristic festival also stands as a dazzling example of how Japanese culture has nimbly adapted its unique affinity for fireflies to fit an ever-changing world.

* * *

In every chapter of this book, we've heard tales illustrating how fireflies' astonishing beauty has been crafted in the creative forge of evolution. But we live in perilous times, here in the Anthropocene. We humans have spread over the Earth, and along the way we've fundamentally altered our local and our global environments. Any creatures that can't keep pace with such changes are destined for extinction. Can you imagine a world without fireflies? I cannot—it breaks my heart to even think about this. Fireflies offer us the gift of wonder, an infallible recipe for falling in love again with nature.

So what can we do to protect fireflies? There are some very simple ways we can each make our local landscapes more firefly friendly (see box). We can also work to preserve and restore those wild places where fireflies thrive—their fields and forests, their mangroves and meadows. During the past few decades our scientific understanding of fireflies' biology and habitat requirements has grown exponentially. This shared knowledge now provides a powerful tool for protecting these silent sparks. We all dream about the kind of world we want our descendants to inhabit. As we contemplate our planetary future, I trust we'll find ways to preserve these stunning ambassadors for Earth's natural magic.

### How Can I Make My Yard More Firefly Friendly?

Here are a few simple ways to make your yard more attractive for local fireflies:

**Create an inviting habitat**

- Let the grass in part of your lawn grow longer by mowing it less frequently. This will help the soil hold more moisture.
- Leave some leaf litter and woody debris in parts of your yard—this makes good habitat for larval fireflies.
- Fireflies need moist places to lay their eggs, so preserve any wetlands, streams, and ponds in your neighborhood.

**Bring back the night**

- When installing or rethinking your outdoor lighting, use only what you need to get the job done.
- Use Dark-Sky compliant, shielded lighting fixtures; these direct light downward, where it's most useful for safety and security. Use bulbs with the lowest possible wattage that will provide just the light you need.
- Turn off outdoor lights when they're not needed, or put them on timers or motion sensors.

**Reduce pesticide use**

- Avoid using broad-spectrum insecticides like malathion and diazinon. Instead, choose horticultural oils or insecticidal bacteria like Bt that will kill specific target pests.
- Get informed about the health and environmental impacts of pesticides. Consider using organic or least-toxic practices and products on your lawn and garden.
- Only apply pesticides when problems arise, never routinely. Don't use Weed & Feed or similar products—they might seem convenient, but they put pesticide when and where it isn't needed.

# A Field Guide to Common Fireflies of North America

* * *

Daily, I'm reminded by a quote from the distinguished naturalist E. O. Wilson that's pinned to my office door, that "mysterious and little known organisms lie within walking distance of where you sit. Splendor awaits in minute proportions." Nowadays we all seem to spend more time connected to our digital devices than to the living world around us. Yet we're innately biophilic—we're irresistibly drawn toward other life forms. So let's pry attention away from our screens and step out into the night.

Earlier chapters of this book related stories of the pleasures, poisons, and plights of many different fireflies from around the world. Now we get to explore firsthand the wondrous world of fireflies. This field guide focuses on the common fireflies of eastern North America. For readers in other parts of the world there are many excellent firefly guides, and these are listed in the notes. During summers across most of the United States, you don't need to travel far to find fireflies—you can just walk out in your backyard or to your neighborhood park. In this section you'll learn how to recognize different kinds of firefly, how to distinguish males from females, and how to eavesdrop on their courtship conversations.

This identification guide covers the five major firefly groups inhabiting eastern North America. We'll get familiar with three different kinds of lightningbug fireflies—*Photinus*, *Photuris*, and *Pyractomena*—and two kinds of daytime dark fireflies, *Ellychnia* and *Lucidota*. Together, these five groups account for the vast majority of fireflies you're likely to encounter.

To recognize different firefly groups, you'll need to look closely at their anatomy as well as their behavior. Although this is a nontechnical identification guide, we'll be using scientific names; only a few fireflies have common names that are widely agreed upon. As you may know, every living organism is scientifically classified by a Latin binomial. In this two-part scientific name, its broader group or *genus* (plural *genera*) comes first, followed by its *species* name. The key and descriptions in this guide are designed to identify firefly genera, though we'll learn to recognize a few species along the way. For each genus, I'll point out its distinguishing features, describe its life cycle, and highlight some notable behaviors.

As a starting point, this guide assumes you have in hand some kind of firefly adult that you want to identify. But how can you tell it's a firefly? That is, does it belong to the family Lampyridae? We learned earlier that all fireflies have bioluminescent larvae, but this key feature is irrelevant when it's an adult beetle that you're trying to identify. It's straightforward to recognize one subset of fireflies, the lightningbugs. These adults possess distinctive glowing lanterns on the underside of their abdomen; even when unlit, these organs appear shiny yellow. A feature common to all adult fireflies is that they're relatively soft-bodied beetles—rather than being shell-like and hard, their wing covers are pliable and leathery. Firefly wing covers, called elytra, are typically black or brown and often have a pale yellow border. Every firefly also carries a wide, flat head shield called the pronotum. Commonly splashed brightly with red, yellow, and black, this firefly pronotum is hard to miss. When a firefly is resting, its pronotum will completely cover its head (though its eyes may stick out a little). Some basic beetle terminology useful for firefly identification is shown in Figure 1.

Below are a few simple ways to distinguish firefly adults from other similar looking, soft-bodied beetles. Just like fireflies, all these beetles have a pronotum, many have similar coloration (black with red or orange markings), and some include pretty good firefly mimics (shown in Figure 7.3).

- In the soldier beetles (family Cantharidae) the pronotum is much smaller, and as a result their heads stick out in front.
- In the net-winged beetles (family Lycidae) the pronotum completely covers the head, but their elytra are sculpted into a ridged, net-like pattern.
- In the giant glow-worm beetles (family Phengodidae), winged males have abbreviated elytra and conspicuous, feathery antennae; the giant, wingless females have pairs of glow spots running down both sides of their bodies.

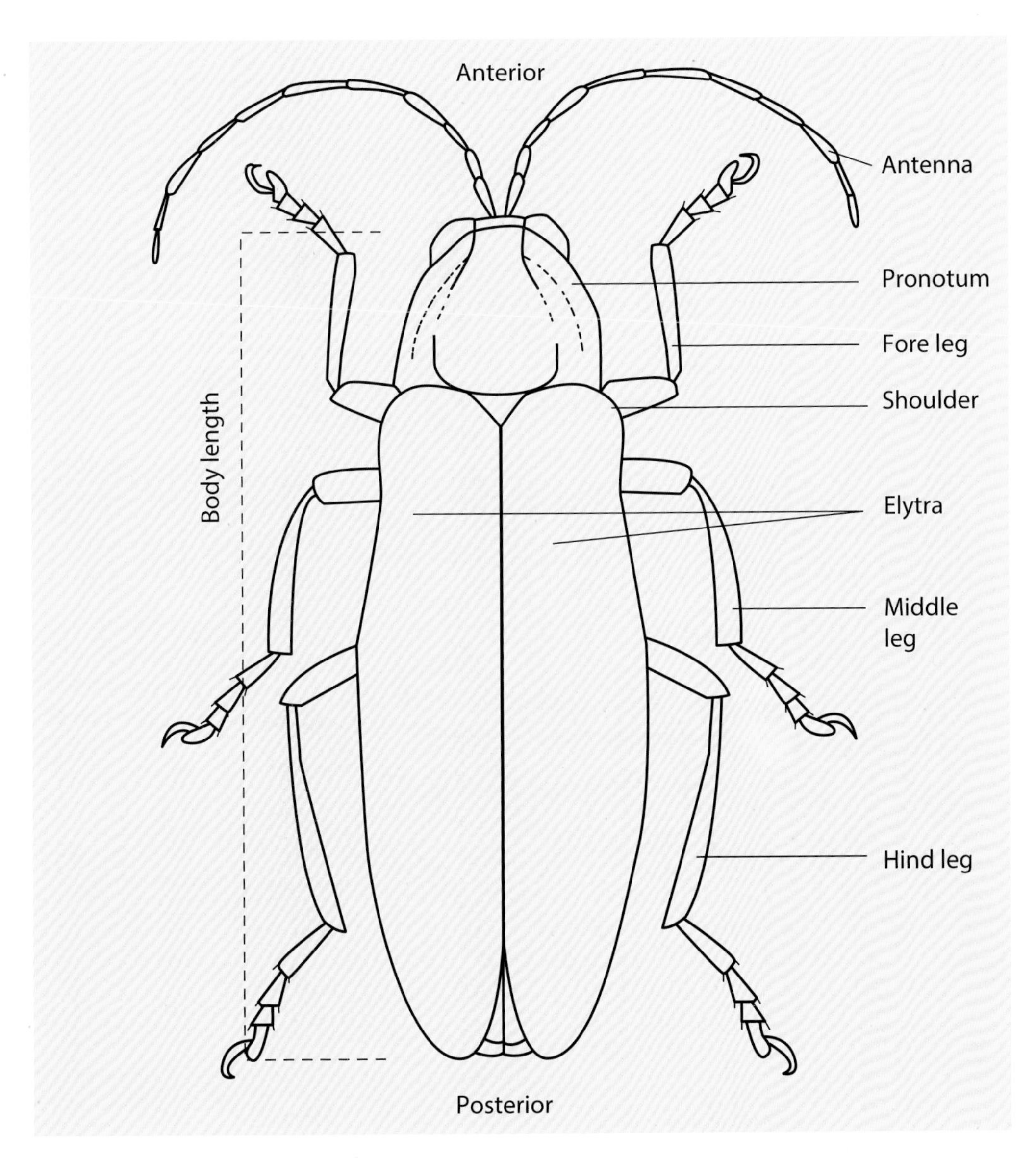


**FIGURE 1** Basic terminology for beetles' external anatomy.

If you're uncertain whether or not your beetle is actually a firefly, the beetle identification guides listed in the Notes section should help.

To examine a live firefly, place it in a clear plastic or glass container and inspect it with a magnifying glass or a 5X to 10X hand lens. If the firefly moves around too much, you can slow it down by placing the container in the refrigerator for a few minutes. Another handy trick is to give the firefly a small bit of apple; while it's busy imbibing apple juice, you can easily check it out.

# Key to Common Firefly Genera

To distinguish among closely related insects, taxonomists typically rely on subtle anatomical details, many of which are only visible when you're examining a dead specimen under a microscope. Here I've chosen to forgo such diagnostic characters. Instead, this key minimizes technical terms and focuses on external features that are relatively easy to see on a living firefly.

Every taxonomic key resembles a treasure hunt. You follow the clues to answer the question: what kind of creature is this? Read each couplet with your firefly (or a photograph of it) in hand, choosing the description that best matches your observations. Once you determine your firefly's genus, you can read more about its life cycle and behavior on the corresponding pages.

**1. Activity period and lantern presence**

a. Adults are active (flying or walking) during evening or nighttime, lanterns are present ........ Go to 2

b. Adults are active (flying or walking) during daytime, lanterns entirely absent, or present but inconspicuous. ........ Go to 4

**2. Pronotum shape**

a. The pronotum has a raised ridge down its midline (Figure 2, *left*; its anterior edge may be slightly pointed ........ Go to *Pyractomena* (p. 155)

b. The pronotum lacks a midline ridge (it may have a slight groove instead; Figure 2, *right*); its anterior edge is rounded. ........ Go to 3

*Pyractomena*: *ridged*

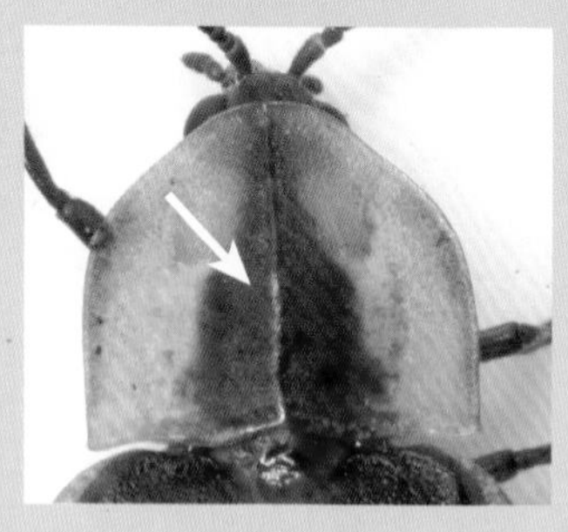

*Photinus*: *grooved*

**FIGURE 2** Pronotum shape: *Left:* In *Pyractomena*, a raised ridge runs down the midline (arrow), and the front margin is slightly pointed (photo by Mike Quinn, TexasEnto.net). *Right*: *Photinus* lacks this midline ridge, but instead often has a shallow midline groove (arrow); the front margin is rounded (photo by Croar.net).

**3. Legs and Shoulders**

a. Legs are long and slender, with hindlegs and midlegs being nearly as long as the elytra; when the shoulder is viewed from the side, the edge of the elytra curves smoothly under, giving the shoulders a hunched appearance (Figure 3, *top*) .......................................................... Go to *Photuris* (p. 158)

b. Legs are short and stout, with hindlegs and midlegs shorter than the elytra; when the shoulder is viewed from the side, the edge of the elytra forms a straight line, making a sharp crease where it folds under (Figure 3, *bottom*) ........................................................................................ Go to *Photinus* (p. 150)

**4. Antennae**

a. Antennae inconspicuous; threadlike and short ................Go to *Ellychnia* (p. 161)

b. Antennae conspicuous; flattened, long and saw-toothed ........................................................................................Go to *Lucidota* (p. 164)

*Photuris*

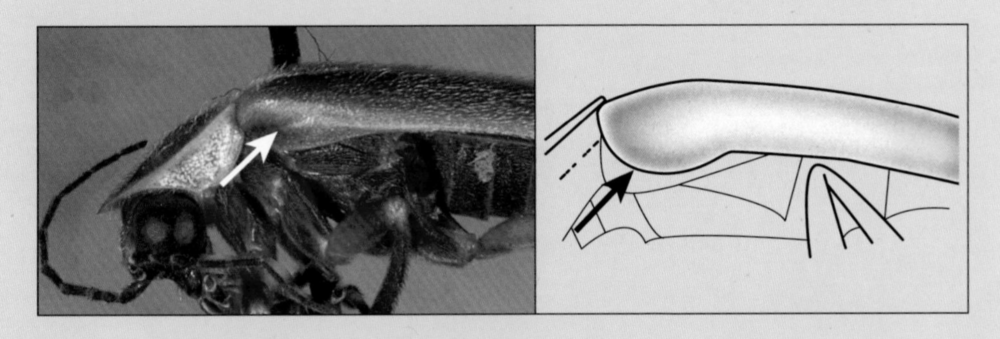

*Photinus*

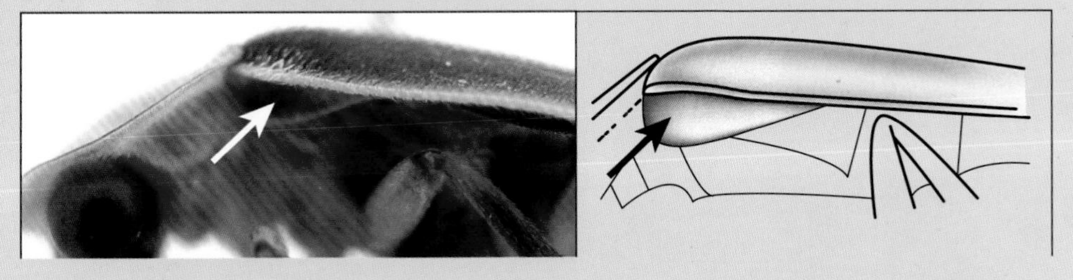

**FIGURE 3** Elytral folds: *Top*: In *Photuris*, the elytral edge curls smoothly under at the shoulder (technically called an "incomplete" elytral fold; photo © Beaty Biodiversity Museum, UBC). *Bottom*: In *Photinus*, the elytral edge continues in a straight line, making a sharp crease where it folds under ("complete" elytral fold; photo by Hadel Go).

# The Lightningbug Fireflies

Lightningbugs are capable of precisely regulating their bright lights to give off quick flashes, which they use during courtship to find mates. Lightningbug fireflies are common in North America, though mainly east of the Rocky Mountains. Across the West, their distribution is restricted to scattered localities in Arizona, Colorado, Nevada, Utah, Idaho, Montana, and British Columbia. Many other western areas may simply be too arid.

There are three major genera of lightningbug fireflies in North America: *Photinus, Pyractomena*, and *Photuris*. At first glance these genera look quite similar; mostly they have elytra that are dark with pale borders, and a pronotum with red, black, and yellow markings. With practice, however, their differences will readily become apparent.

## *PHOTINUS*

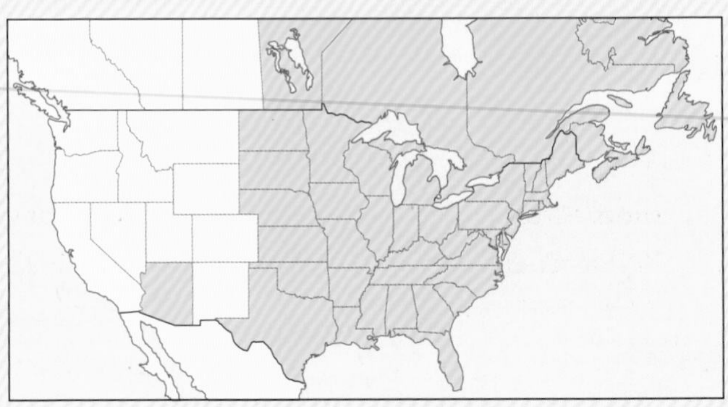

### Anatomy

*pronotum*—flat, though sometimes with slight groove down the midline; front margin is rounded, side margins are yellow (infrequently dark); central pink area is typically interrupted by a wide black stripe or spot.

*body*—body length 6–15 mm, slender; legs short.

*elytra*—usually black (rarely gray) with yellow borders; side margins parallel; when the shoulder is viewed from the side, the edge of the elytra forms a straight line, making a sharp crease where it folds under (Figure 3, *bottom*).

There are 34+ species of *Photinus* fireflies in the eastern United States and Canada, distinguished from one another mainly by the shape of their male genitalia (Green 1956) and secondarily by their flash pattern (Lloyd 1966). *Photinus* are the familiar fireflies that fill our summer evenings with delight—easy to chase, they fly low above the ground at a leisurely pace. Beginning at twilight, these lightningbugs will continue to flash for an hour or two after sunset. Although adults of any particular species stay active for only a few weeks, collectively *Photinus* will fly all summer long; as soon as the mating season of one species ends, another one begins.

### Sexual dimorphism

*Photinus* males are easily distinguished from their females: in males, the light-producing lantern occupies the entirety of the last two abdominal segments (Figure 4, *left*), while the female lantern is restricted to a much smaller area in the middle of the penultimate segment (Figure 4, *right*). (In both sexes, there may be pale but nonluminous areas adjacent to the lantern.) Males' eyes are also much larger than those of females. Finally, although most females have wings just like their males, females in a few *Photinus* species have abbreviated or no wings.

### Life cycle

The *Photinus* life cycle begins when a female deposits some eggs in moist soil or moss. After about two weeks, tiny glowing larvae hatch out. *Photinus* larvae live belowground, feeding mainly on earthworms and other soft-bodied insects. In captivity they feed gregariously, but it's not known whether they do so in nature. In northern latitudes, *Photinus* fireflies probably spend between one and three years as larvae; farther south, larvae can complete their development within a

**FIGURE 4** *Photinus* sexual dimorphism: *Left*: Male lantern fills two abdominal segments, and their eyes are larger compared to females (photo by Terry Priest). *Right*: Female lantern is restricted to the middle of the penultimate segment (photo by Andrew Williams).

few months of hatching. In late spring each larva constructs an igloo-like soil chamber and curls up inside. Within a few days it transforms into a pupa, then emerges about three weeks later having morphed into an adult firefly.

### COURTSHIP

*Photinus* species are very habitat specific: some species prefer open grasslands, while others are found under the forest canopy, along rivers, or in freshwater marshes. When several species occupy the same habitat they court in shifts, partitioning the night into distinct activity periods. Males of twilight species begin flying at dusk and may stay airborne for only twenty to forty minutes. In shady habitats or on cloudy days, such males may start flying even before sunset. In other species, males wait until after dark to begin their courtship flights, then fly for one to two hours.

During their courtship period, male *Photinus* fly slowly and typically stay less than 2 meters above the ground. As they fly, males advertise their availability by broadcasting a flash pattern characteristic for their species. Although most females have fully developed wings and *can* fly, they rarely do so. Instead they take up perches on grass or low vegetation to admire the passing males. When a female *Photinus* is interested in a male, she'll respond by flashing back to him after a short delay. He flashes again, and they strike up a conversation. This back-and-forth flash dialogue sometime continues for hours, until finally they meet and mate. Chapter 3 explored *Photinus* courtship rituals in intimate detail.

### PHOTINUS COURTSHIP CODES

Their predictable flash dialogues make it relatively simple to decipher the courtship conversations of *Photinus* fireflies. Our knowledge of firefly linguistics comes mainly from studies done by Jim Lloyd, the firefly biologist we met in chapter 3. Lloyd spent the 1960s traveling around the eastern United States, carefully recording male flash patterns and female flash responses for about two dozen different *Photinus* species (Lloyd 1966). He also brought along a thermometer, because air temperature alters the timing of fireflies' signals, just like it does with other cold-blooded creatures such as crickets, frogs, and katydids.

As the flash chart shown in Figure 5 indicates, some *Photinus* males court using a single pulse of light that they repeat at regular intervals. This single pulse can be quick and snappy; for example, in *Photinus sabulosus* it lasts a mere 1/10th of a second. Other times it's more prolonged, as in *Photinus pyralis*: these males stay lit

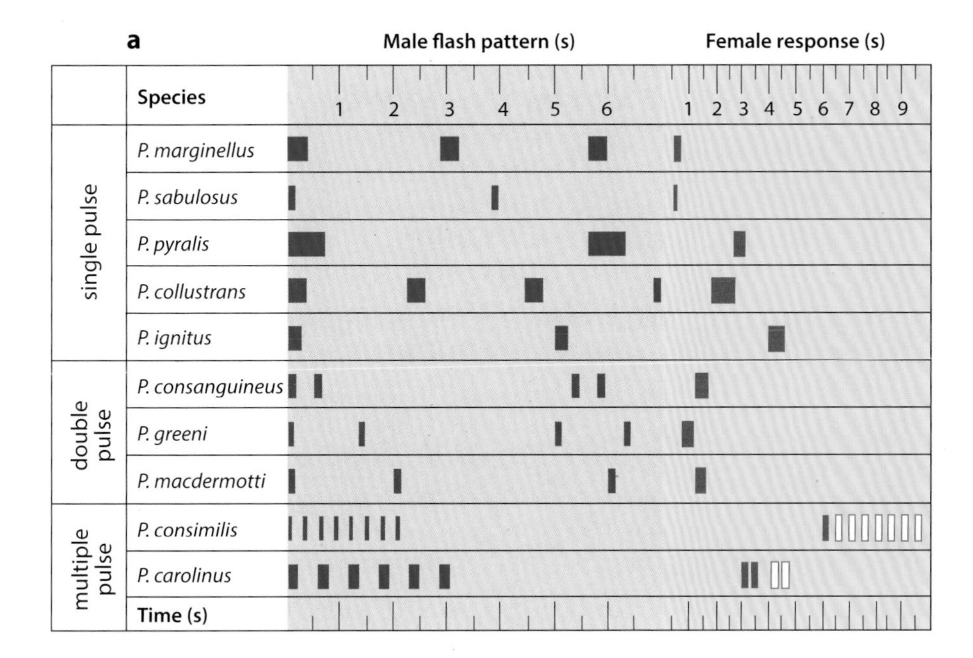


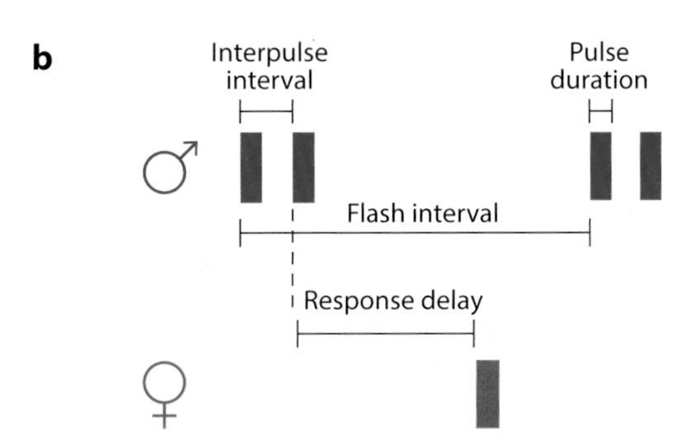


**FIGURE 5** (a) Courtship flash patterns used by different *Photinus* species. Time scale at top indicates seconds, rows are different *Photinus* species. Males' flash pattern is shown on left (in blue), female responses on right (in red). Open boxes indicate optional flashes (modified from Lewis and Cratsley 2008). (b) Terminology for flash patterns: note that female response is timed from the beginning of the final pulse in the male's signal.

for about three-fourths of a second. Males of some other *Photinus* species emit a pair of pulses (as in *Photinus consanguineus*, *greeni*, and *macdermotti*), while still others emit a train of multiple pulses (as in *Photinus consimilis* and *carolinus*). Again, males repeat their characteristic flash patterns at regular intervals as they fly.

How do females recognize the males of their own species? By testing females with simulated flashes, Lloyd discovered that females pay attention to the males' flash timing. In single-pulse species, females use mainly pulse duration and flash interval to recognize their own males. In double- or multiple-pulsed species, females pay close attention to the number of pulses and the interpulse interval.

Female responses are illustrated on the right-hand side of the flash chart, where you'll see that most females use a single-pulsed response flash. But when females of certain species become enthusiastic, their reply sometimes contains several pulses. Photinus *consimilis* females, for example, may give anywhere up to twelve. Lloyd discovered that male *Photinus* recognize their own females according to how long the female waits before she starts to flash. This so-called female response delay differs among species. Females of some species answer after a very short delay (less than a second), while others wait much longer before responding. Females of *Photinus ignitus*, for instance, wait four or more seconds before they answer a courting male.

It's important to recognize that because firefly flash signals are temperature dependent, the flash chart shows only approximate timings for each species. It will be pretty accurate for air temperatures between 66° and 75°F (19–24°C). In warmer temperatures, things will speed up: males' pulse duration, interpulse interval, and flash interval all get shorter, as will the females' response delay. Conversely, at cooler air temperatures everything will slow down. For instance, males of *Photinus pyralis* fireflies flash once every 5.5 seconds at 75°F, but slow down to once every 8 seconds at 65°F. Once you know the correct courtship code for your local *Photinus* species, it's easy to start engaging them in conversation using just a penlight, as explained in the first *Stepping Out* adventure below.

There's one *Photinus* firefly that deserves special mention. Also known as the Big Dipper, *Photinus pyralis* fireflies are often seen in city parks, suburban lawns, grassy fields, and along roadsides. This common name derives from a combination of their large size (they can reach more than 1 centimeter in body length) coupled with these males' distinctive dip-and-flash habit. At dusk, *Photinus pyralis* males fly slowly in search of females, emitting a half-second flash about every six seconds. As the male starts to flash, he first dips down then rises up sharply, repeatedly skywriting the letter *J* as he goes. After each flash, the male hovers momentarily, watching for a female's reply. Big Dipper females remain perched in the grass and respond to males with a single flash delivered at a 2- to 3-second delay. Big Dipper fireflies have been studied extensively by physiologists and biochemists. For many decades, these fireflies were commercially harvested to extract their light-producing chemicals, as related in chapter 8. Fortunately, in spite of this collecting pressure, Big Dipper fireflies remain common throughout the eastern United States.

# *PYRACTOMENA*

*Pyractomena angulata*

### ANATOMY

*pronotum*—sculpted (not flat), with side margins that flare upward and a raised midline ridge; front margin often slightly pointed; side margins are dark in many (not all) species.

*body*—shape varies from broad (as in *Pyractomena angulata* pictured in the upper photo) to more elongate (as in *Pyractomena borealis*, lower photo); body length 7–22 mm; legs short.

*elytra*—typically black with yellow borders.

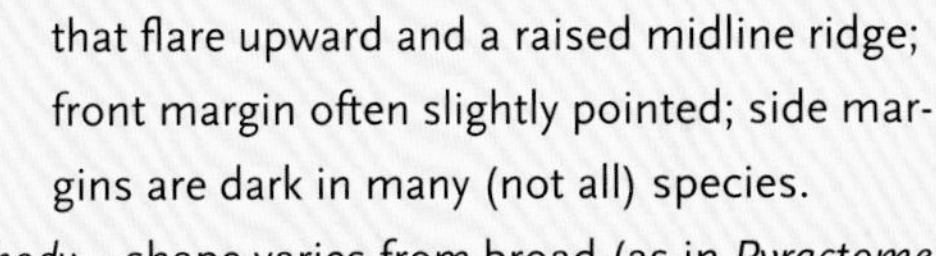

*Pyractomena borealis*

North America hosts sixteen species of *Pyractomena*, including several species that extend into Colorado, Utah, and other western states. They can be distinguished from other lightningbugs by the features described above. Within *Pyractomena*, different species are distinguished by their general body shape, by the shape of the males' genitalia, and by the pattern of fine hairs on their elytra (Green 1957).

## Sexual dimorphism

As with *Photinus*, the sex of a *Pyractomena* firefly can be determined by the shape of its light-producing lantern. In males, this lantern takes up the entirety of the last two abdominal segments (Figure 6, *left*). In females, the lantern is restricted to four small spots at the sides of these segments (Figure 6, *right*). (In both sexes, there is often some pale, but nonluminous tissue, adjacent to the lantern itself. One way to see what glows and what doesn't is to bring the firefly into a dark room and gently disturb it.) Males also have much larger eyes than females. In *Pyractomena,* females are never wingless.

## Life cycle

*Pyractomena* fireflies are often found in wet meadows, woods, marshes, and along streams. Their larvae glow, as do all firefly larvae, and snails are their preferred food; their elongated, wedge-shaped heads fit neatly inside snail shells. Some *Pyractomena* larvae are semiaquatic, foraging both above and underwater. In contrast to most other fireflies, *Pyractomena* larvae pupate above ground, crawling up onto vegetation.

One species, *Pyractomena borealis*, is found in forested habitats (photo above). This species is easily recognized by its large size (ranging from 11 to 22 mm body length), dark body coloration, and narrow yellow borders on its elytra. Widespread across eastern North America, this lightningbug occurs from Maine through Wisconsin and south to Florida and Texas in the United States, and spans Canada from Nova Scotia to Alberta. In southern portions of its range, larvae climb tree trunks in late winter looking for a sun-warmed place to pupate (Figure 7). Farther

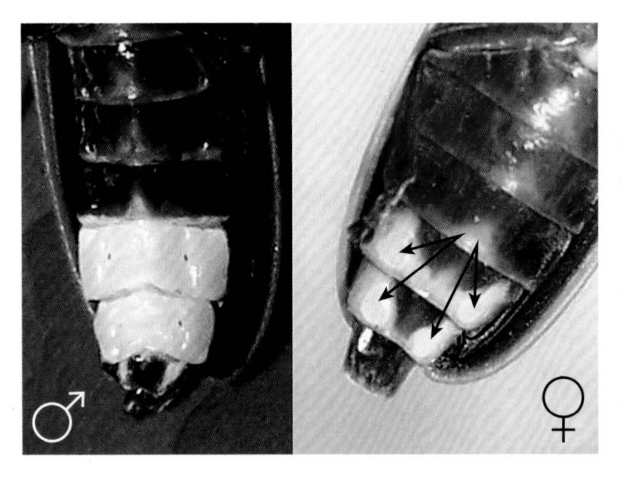


**FIGURE 6** *Pyractomena* sexual dimorphism: *Left*: Male lantern fills the entire width of two abdominal segments. *Right*: Female lantern consists of four small spots (arrows) at the sides of these segments. Note the nonluminous, pale yellow areas surrounding these spots (*Pyractomena borealis* photos by Lynn Faust, from Faust 2012).

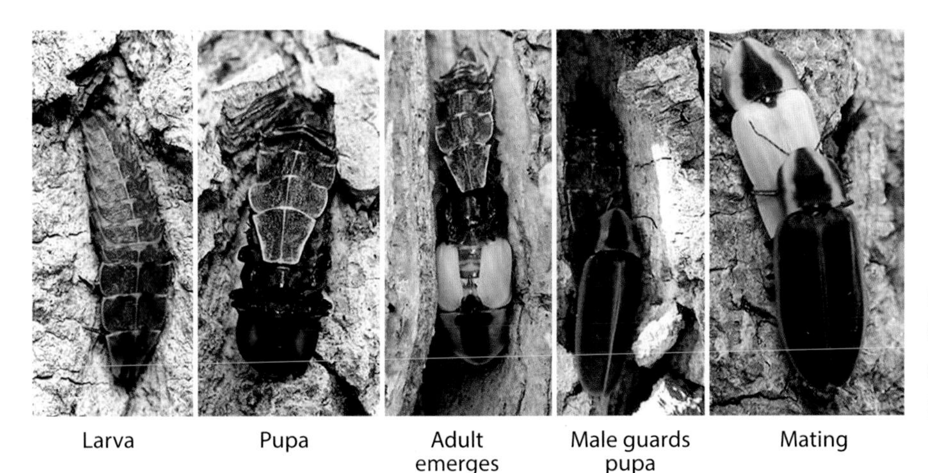


**FIGURE 7** Five stages during the life of a *Pyractomena borealis* firefly (photos by Lynn Faust, from Faust 2012).

north, these larvae overwinter on tree trunks and pupate in the early spring. When the adults emerge from their black pupal cases 1–3 weeks later, their bodies are initially soft and white; a day later, they've hardened and become pigmented.

## COURTSHIP

Like *Photinus*, *Pyractomena* adults locate their mates using call-and-response flash dialogues, and their flash timing is also temperature dependent. Tolerant of cold weather, *Pyractomena borealis* are often the first lightningbug fireflies to appear in springtime. In Florida they can be seen flashing in treetops in late February; in Tennessee they appear in late March through April, and farther north their mating season begins in May–June.

In *Pyractomena borealis*, males emerge first, and they immediately start crawling along the tree trunk searching for females. Males will stand guard over an immature female, waiting to mate as soon as she emerges (Figure 7; Faust 2012). When males begin to fly, they patrol high in the treetops starting an hour or so after sunset, searching for females. These males emit a single, short flash every 2–4 seconds, depending on air temperature; females perch on tree trunks and respond with a single, short (1/2 second) flash delivered at a 1-second delay.

Another distinctive species is *Pyractomena angulata*, whose males are easy to recognize by their flickering, amber-colored flashes that last about a second. One of my favorite fireflies, they look like flickering candle flames. Flying over marshy grasslands, shrubs, and even up into trees, males repeat these flashes every 2–4 seconds. Seen up close, this is the widest-bodied of all *Pyractomena* species, its elytra having broad yellow side borders.

# *PHOTURIS*

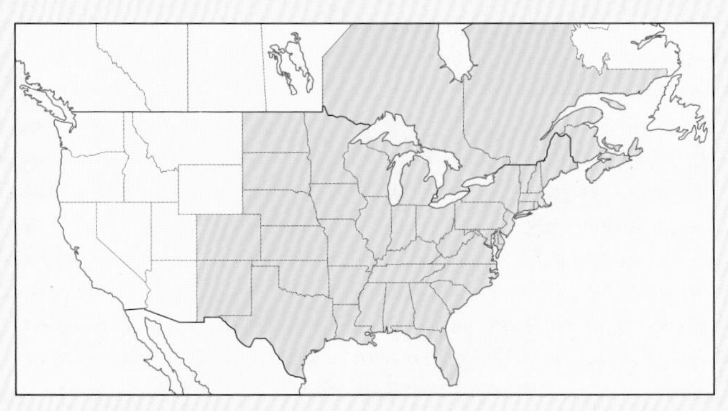

### Anatomy

*pronotum*—head projects out slightly in front when the beetle is walking; semicircular shape, side margins are yellow (not dark); typically decorated with a central red spot and black markings.

*body*—large (10–20 mm body length), long-legged, with midlegs and hindlegs nearly as long as the elytra; hunch-backed appearance; body ellipsoid, side margins not quite parallel.

*elytra*—black or brown with yellow borders; pale lines often extend diagonally from each shoulder; when the shoulder is viewed from the side, the elytral margin curls smoothly under (Figure 3, *top*).

Twenty-two species of these hunchbacked, long-legged, and agile lightningbugs have been described in North America, but some estimates put the true number closer to fifty. Adult females of many *Photuris* species (though not all) specialize in hunting other lightningbugs. Telling different *Photuris* species apart is quite challenging, even for Jim Lloyd, an expert who been studying this confusing firefly group for forty years. The first problem is that the genitalia of all *Photuris* males look virtually identical, rendering this standard feature for distinguishing insect species worthless. The second problem comes from the tremendous versatility of their flash behavior; a single species can produce a dizzying array of flash patterns.

Males of many *Photuris* species give different flash patterns depending on the time of night or what other fireflies are nearby (Barber 1951). And *Photuris* females readily switch between giving their own courtship flashes and imitating their prey's flash signals. In spite of these difficulties, *Photuris* fireflies as a group are easy to distinguish from other lightningbugs by the features described above.

### Sexual dimorphism

While initially somewhat tricky, with practice it's easy to determine the sex of a *Photuris* firefly based on the size and shape of its lantern. The male's lantern completely covers his last two abdominal segments, as in the two previous groups of lightningbugs. In females the lantern is on the same two segments, but it doesn't quite reach all the way to the edge; instead, it is surrounded by a pale border of nonluminous tissue (Figure 8).

### Life cycle

*Photuris* larvae are frequently seen at night, glowing dimly and crawling at the surface of damp roadsides, paths, and lawns. Dietary generalists, these larvae are omnivores and scavengers, consuming snails, worms, soft-bodied insects, even berries. When ready to pupate, they congregate in small groups, and each larva builds a small earthen chamber. The adults emerge after 1–3 weeks. Most *Photuris* females are unusual in their eating habits; while other adult fireflies eat little or nothing, these females specialize in preying on other fireflies. They mainly attack males of *Photinus*, but also consume *Pyractomena* and occasionally other *Photuris*. As described in chapter 7, these females gain access not only to their victims' protein but also to their poisons. *Photuris* females accumulate their prey's defensive chem-

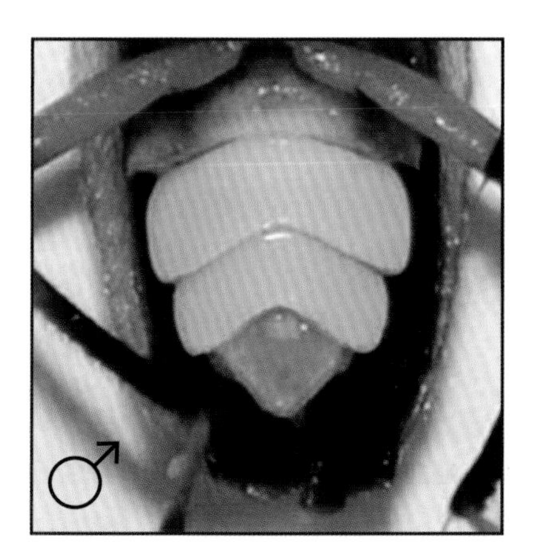


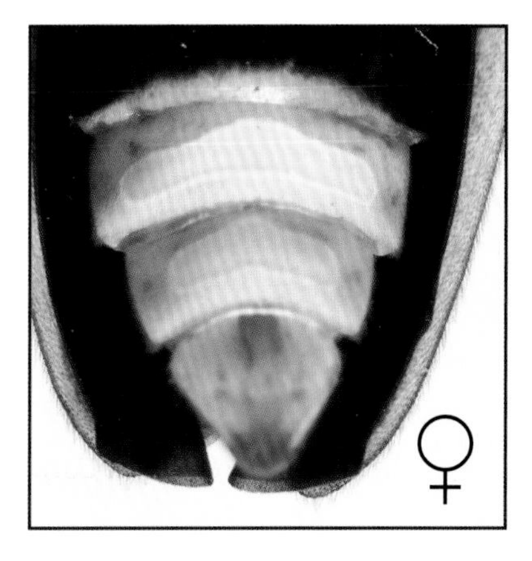


**FIGURE 8** *Photuris* sexual dimorphism: *Left*: Male lantern fills two abdominal segments. *Right*: Female lantern occupies only the middle part of the same two segments; it does not extend fully to the edges. Note the nonluminous areas surrounding the female lantern (photos by Rebecca Forkner 2010, Marie Schmidt).

icals, which they use to protect themselves and their eggs. *Photuris* adults are relatively long-lived—in captivity they can survive for a month or more.

### Behavior

These lightningbugs are fast flying and agile; when caught, *Photuris* easily escape from your net or your hand. In general, *Photuris* fireflies fly later during the night and higher off the ground compared to *Photinus.* When you wake in the middle of the night to find a firefly flashing frenetically as it walks across your window screen, it's most likely *Photuris*.

*Photuris* females lurk within other fireflies' courtship aggregations and capture prey using several different hunting tactics. Femmes fatales lure prey males close by deceptively imitating the flash responses produced by the females of their target prey. They also chase and kill prey males in flight and wait near spider webs to steal firefly prey that get trapped. These specialist predators are a prime agent of natural selection for other North American fireflies.

In several species, males switch between two different flash patterns depending on the time of night. For example, males of *Photuris tremulans* start off by giving flickering, long (1 second) flashes, then switch to giving quick, single flashes. Both sexes also flash erratically in contexts other than courtship, such as when they're landing, taking off, or walking. This impressive versatility makes using *Photuris* flash patterns for species identifications very difficult. Because *Photuris* often mate high up in trees, very little is known about their mating behavior.

## The Dark Fireflies

Adults of many fireflies don't light up. These day-active, dark fireflies can be found coast to coast across the United States and Canada. While some people might not consider these to be "real" fireflies, their genetic similarity, bioluminescent larvae, and other shared features do indeed make these dark fireflies authentic, card-carrying members of the family Lampyridae. *Ellychnia* fireflies, described below, are close relatives of *Photinus* lightningbugs, even though they've branched out to adopt a very different lifestyle. Dark fireflies also include two *Photinus* species whose adults lost their light-producing ability even more recently. It may be that these daytime dark fireflies have shifted their activity to evade nocturnal hunters like the predatory *Photuris* fireflies. While day-active fireflies are presumed to use airborne scents to find and attract their mates, the exact chemicals involved have yet to be identified.

## *ELLYCHNIA*

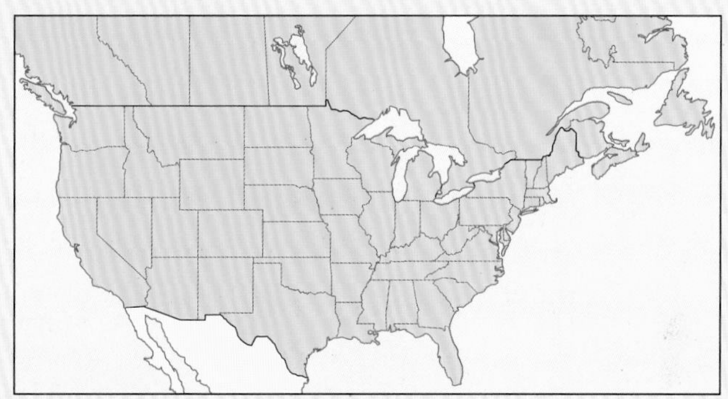

### ANATOMY

*pronotum*—semicircular shape; side margin typically are dark, although yellow and red coloration varies between species.

*body*—broad bodied, with length ranging from 6 to 16 mm; lantern absent; legs short and stout.

*elytra*—dark olive to jet black, lacking pale borders; sometimes with lengthwise costae (ridges) that can be more or less pronounced.

Close cousins to *Photinus*, this group of daytime dark fireflies consists of twelve+ nonluminous species. *Ellychnia* are widely distributed throughout North America, with several species exclusively found west of the Rocky Mountains. Three species "complexes" occur in the eastern United States, although distinctions among these remain muddled (Fender 1970). *Ellychnia corrusca* (pictured above) is the most ubiquitous eastern species. Markings on its pronotum make it easy to recognize: it has a large black spot in the center that's bordered with red and enclosed within two pale areas shaped like a set of parentheses. Sometimes called Winter fireflies, these are among the first insects to become active each spring, when adults can be seen flying slowly through forested areas.

### Sexual dimorphism

Because they lack lanterns, distinguishing between females and males in *Ellychnia* requires closely examining the underside of the beetle's abdomen. In females, the terminal segment is triangular and has a small central notch (Figure 9, *right*); in males, this last segment is much smaller, rounded, and unnotched (Figure 9, *left*).

### Life cycle

*Ellychnia* larvae, like those of all other fireflies, are bioluminescent and carnivorous. Rarely seen, they live and hunt within decaying wood.

*Ellychnia corrusca* ranges from all the way from Florida north to Ontario, and its life cycle varies with latitude. In the northern part of its geographic range, these adults emerge in autumn. They crawl up onto tree trunks, where they wedge themselves into grooves and hunker down to spend the winter. Decidedly

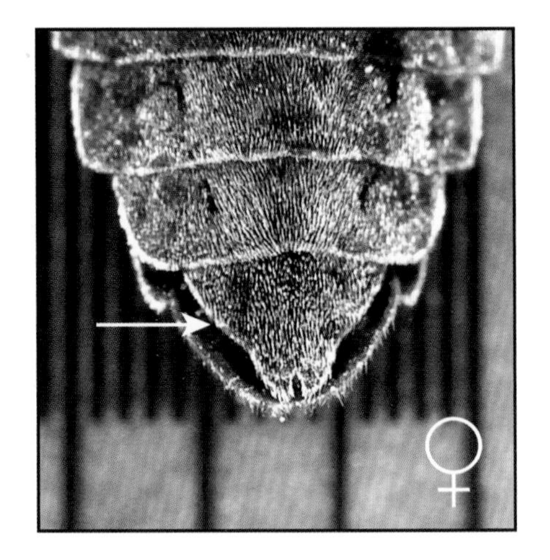


**FIGURE 9** *Ellychnia* sexual dimorphism: *Left*: In males, the last abdominal segment is rounded and small. *Right*: In females, the last segment is triangular, larger, and notched at its tip. Use a hand lens to look at the beetle's underside (ruler shows millimeters).

hardy, the adults remain inactive through months of subfreezing temperatures. Our mark-recapture study estimated that about 90% of these adults manage to survive the winter in Massachusetts (Rooney and Lewis 2000). We often found beetles, apparently dead, lying on their backs in the snow; these quickly revived in the warmth of my car.

Mating takes place in early spring (March and April), and females lay their eggs nearby. Hatching out in early summer, larvae spend the next sixteen months eating and growing. Not until late summer of their second year will they pupate in rotting logs. Then, in the fall, the adults emerge to overwinter on tree trunks again.

Farther south (below about 40 degrees north latitude), their life cycle follows a very different pattern (Faust 2012). Here adults emerge in late winter (late February and March) and immediately start crawling up tree trunks. They mate in early spring (March and early April), as in northern climes. But southern larvae can complete their development during a single summer and fall; they're ready to pupate by late fall, with adults emerging in late winter. Thus, in southern regions, it's warm enough for *Ellychnia corrusca* to complete its entire life cycle within a single year. With global climate change, higher temperatures will likely speed up development and allow even northern populations to complete their life cycle within one year.

### Behavior

These nonluminous fireflies are most conspicuous in early spring, when the adults start crawling up tree trunks, mating, and flying through forested habitats. Mating generally occurs on tree trunks or on the ground, where pairs can be seen in their tail-to-tail mating position for twelve hours or more. When the adults emerge—early fall in the north, late winter in the south—dozens may be found congregating on particular trees. They prefer large trees with deeply grooved bark, and they frequent the same trees year after year. During springtime adults are attracted to maple sap, and they commonly end up in sap-collecting buckets.

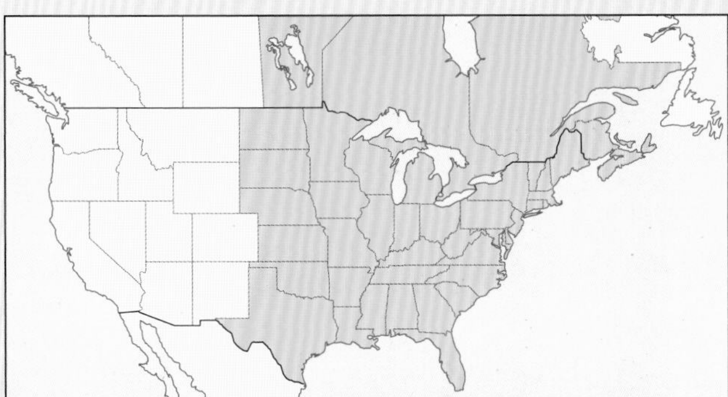

### Anatomy

*pronotum*—variable in both shape (apex rounded or pointed) and coloration (from all red to yellow or black with two red spots).

*body*—broad body ranging from 6 to 14 mm in length; lanterns are absent or vestigial; antennae are long, flat, and saw-toothed; the second antennal segment (counting from the eyes) is tiny, much shorter than all other segments although equal in width.

*elytra*—matte black, without pale borders.

The three North American species of day-flying *Lucidota* fireflies have distinctive flat, saw-toothed antennae. These adults rely on chemical signals to find their mates. By far the most abundant species is *Lucidota atra* (pictured above), common during midsummer throughout the eastern United States. Often seen flying slowly within a few meters of the ground, these large adults (7.5 to 14 mm) are stunning: jet black with a yellow pronotum whose central black stripe is flanked by two bright red spots. Adults may have vestigial lanterns, which appear as small pale spots located on the last one or two abdominal segments of females and males, respectively.

### Sexual dimorphism

Males' antennae are thicker, longer, and more serrated than those of females (Figure 10).

**FIGURE 10** *Lucidota* sexual dimorphism: Males' antennae (on *right*) are flatter, more saw-toothed, and conspicuous compared to those of females (on *left*; photo by Molly Jacobson).

## Life Cycle

*Lucidota* larvae live within and under decaying logs, where they feed on snails, worms, and soft-bodied insects. The nearly grown larvae or pupae become inactive during winter, then metamorphose into adults in early to midsummer. Adults fly during the daytime, and are commonly seen flying above lawns and grassy areas, through marshes, and along streams and forest edges. In one species (*Lucidota luteicollis*), adult females are wingless and can't fly.

## Behavior

Some daytime dark fireflies, including *Lucidota atra*, have been shown to use airborne perfumes to find their mates (Lloyd 1972). Their stationary females give off chemical signals that get carried on the wind; known as pheromones, these chemicals create an invisible plume that extends downwind from each female. Males search for females by flying back and forth until they detect a plume, then travel upwind to locate the female. The chemical compounds released by females haven't yet been identified, although the fact that males are only attracted to females of their own species suggests these olfactory signals are species-specific.

# Stepping Out—Further Firefly Adventures

* * *

Many people will be content just to identify the fireflies they encounter. Yet, like bird watching, a breathtaking world of wonder awaits anyone seeking further firefly adventures. Here are a few suggestions that may inspire you to step out and explore the unseen world of fireflies (check the book blog for others).

## Talking to Fireflies

Light is the language of love for lightningbug fireflies, and their visible call-and-response dialogues make it easy to eavesdrop on their courtship conversations. Once you decipher the code, you can even grab a penlight and talk to your local fireflies. If you can find them, the courtship conversations of *Photinus* lightning-bugs are easiest to join.

Start by watching and learning the males' flash pattern. Stand or sit quietly in the fireflies' habitat as their flight period begins. If you're lucky, you'll see the males lifting off to begin their courtship missions. Focusing on a single flying male, time his flash pattern with a stopwatch or by counting off seconds (*one, one thousand, two, one thousand . . .*). For dusk-flying species—*Photinus pyralis*, for instance—you can still make out their tiny bodies, so it's easy to follow a single male. For species that fly later, the trick I use to keep track of individuals is to crouch below them so their bodies become silhouetted against the sky. After timing several males, you'll be able to pick up their flash rhythm. Using a penlight (see the list of field gear for recommendations), try to imitate the males' flash pattern. It will take some practice, but soon you'll feel just like one of the crowd.

Now you're ready to search for female *Photinus*. Because females respond infrequently, they're more difficult to find. Watch to see where the males are searching and, using your penlight like a fishing lure, start trolling for females. Walking slowly, keep repeating the male flash pattern. You can hold your finger over the tip of the light so it doesn't shine too brightly. You don't need to worry about the flash gesture or the exact color of the light; studies have shown that females don't respond to flash shape, and fireflies don't see in color.

After each flash, be alert for any answer coming from a female, who'll be perched down in the grass or on low vegetation. Males may also be flashing from the ground, but they'll be walking around; in contrast, females generally stay in one spot, motionless. If a female is interested, she'll respond to you or to a passing male with a single flash (see Figure 5). Such female flashes are typically longer lasting than those of males, and their intensity noticeably rises then falls. When you receive an answer, move toward the female and flash again. Keep in mind that you're competing against firefly males who are desperately seeking mates, so you'll have to be quick! Males generally outnumber females, and they will quickly locate any responsive females.

You can also try searching for females in spots that males are likely to miss—for instance, around bushes or under trees at the habitat's edge. Another strategy for finding females is to wait until all the males have stopped flying. After the flight period has ended, you'll still see clusters of flashes near the ground. These are ongoing courtship dialogues—typically several males are competing to locate a single female. You can sometimes find females by looking for these flash clusters, which may last long into the night.

Once you find a female, get as close as possible but be careful not to knock her off her perch. Measure her response delay with a stopwatch or by counting; for males with double- or multiple-pulsed flash patterns, start timing from the beginning of the male's final pulse. You'll get a more accurate estimate by measuring the female's response delay several times. It's a good idea to turn on your headlamp briefly to make certain it's actually a *Photinus* female that you've been watching, and not an imposter femme fatale.

At last you're ready to join the conversation! To imitate a female, hold the tip of your penlight against your finger (again, this dims the light) and place it near the ground. Focus on a single male. When he flashes, count out the correct delay before flashing back your answer. When he comes closer and flashes again, keep flashing back to him with the proper response. Don't get too enthusiastic, though; keep in mind that the firefly females you're imitating are coy, and they never answer every male flash. By imitating a female, you should attract quite a bit of male attention! I've drawn in firefly males from quite a distance, and sometimes I've had as many as a dozen male fireflies land along my arm and penlight.

You've already seen how to locate females by imitating the males' flash pattern. A marvelous thing happens when you do this near the end of the mating season. By then, as explained in chapter 3, available females often outnumber willing males. Walk out to the middle of the firefly habitat and imitate a male's flash pattern with your penlight. If you're lucky, a chorus of females will answer you, as they all flash back with identical time delays.

If you're interested in knowing exactly what species of *Photinus* fireflies are active in your neighborhood, get started by recording both sexes' flash patterns. For males, note the number of pulses and use a stopwatch to time the males' interflash intervals. If the male flash pattern contains two or more pulses, measure the interpulse interval. Don't worry about pulse duration, which requires specialized recording equipment; just note whether the flash is either quick (less than one-half a second) or more prolonged (more than one-half a second). For females, measure their response delay. You'll also need to record the air temperature, because this affects flash timing. It's best to collect this information over several different nights, keeping track of your observations with a notebook or voice recorder.

Referring to the flash chart in this section, find the pattern that resembles your observations most closely. Keep in mind that everything will slow down at cooler temperatures, and speed up at warmer ones. Jim Lloyd's 1966 monograph on *Photinus* fireflies provides much additional information, including range maps, habitat descriptions, activity periods, and flash behavior for more than two dozen US species. This monograph can be downloaded from the University of Michigan for free, using the link provided in the reference section.

## The Invisible World of Firefly Perfumes

With their bright lights and flashy displays, lightningbugs have attracted lots of scientific attention, as well as popular acclaim. But what about all those daytime dark fireflies—lacking lights, how do they manage to find their mates? While only a few scientific studies have explored the courtship rituals of these poor cousins, evidence suggests they rely on invisible chemical signals that waft through the air.

One classic study looked at *Lucidota atra*, large, black day-flying fireflies that are easy to spot in midsummer as males fly slowly through forests, across lawns, and along roadsides. To see whether these females attract mates by releasing airborne perfumes, Jim Lloyd did some simple field experiments on Michigan's Upper Peninsula (Lloyd 1972). He placed females into small, shallow containers (these were 10-centimeter-wide petri dishes, a workhorse in scientific laboratories). Lloyd carefully covered each dish with mesh fabric; this kept the females in (and males out), while still letting air circulate freely.

He set these captive females on the ground in a woodland clearing. Then he waited and watched. A light breeze was blowing, and within three minutes males started arriving from downwind and landing directly on the females' dishes or nearby. Lloyd counted thirty males who flew in within the first half hour. He also marked some males and then released them at various distances downwind from the captive females. When he released males from 9 meters away (30 feet), every male found the female within eight minutes. One speedy male, marked and released from 27 meters (90 feet) away, managed to locate the female in just over three minutes! These females were releasing chemical signals that were carried downwind, forming a diffuse, invisible plume. Males seem to locate their mates by flying around until they smell a female, then flying upwind to find her.

Research done on another day-active firefly also supports the idea that females are releasing perfumes to attract males. Raphaël De Cock, the Belgian musician-scientist introduced in chapter 5, studied *Phosphaenus hemipterus*, the lesser European glow-worm (De Cock and Matthysen 2005). Working on the grounds of the University of Antwerp, Raphaël ran some experiments to find out if *Phosphaenus* females might also attract males using airborne chemical cues. When he placed females in gauze-covered dishes, they attracted nearly thirty males within an hour, the majority arriving from downwind. Raphaël also described females making a "calling" gesture, curling their abdomen sideways as they released their perfume.

And that's pretty much all we know about how daytime fireflies find their mates—this field is wide open. Not only does the nature of firefly perfumes remain a mystery, but also mate-finding behavior in most species hasn't even been described. Daytime fireflies are found worldwide. One that's common throughout the eastern United States is *Ellychnia corrusca*; in New England, their mating season lasts several weeks during April and May. Yet we know hardly anything about their courtship behavior. And the western United States is home to many other daytime dark fireflies. So anybody who's willing to observe and experiment should be able to discover something entirely new!

Here's one experiment you can easily try. Gather several shallow dishes (Mason jar lids would work well), some mesh fabric (such as mosquito netting or tulle from a fabric store), rubber bands, and a sponge. Collect some of your local daytime dark fireflies, and use the identification guide to separate them by sex. Place a small piece of moist sponge in each dish to provide moisture. Set up several dishes with females, as well as some identical ones with males, and others that are empty (Figure 11). The empty dishes will help rule out the possibility that males might simply be attracted to the dish setup; the captive males will show whether males get attracted to others of their species, regardless of sex.

Place the dishes in groups around your fireflies' habitat, spacing them several meters apart. You can put each dish on a white plate or cardboard circle to make it easier to detect arriving males. Settle down to keep watch, or check periodically to count any arrivals. Are more males attracted to the dishes containing females than to the others?

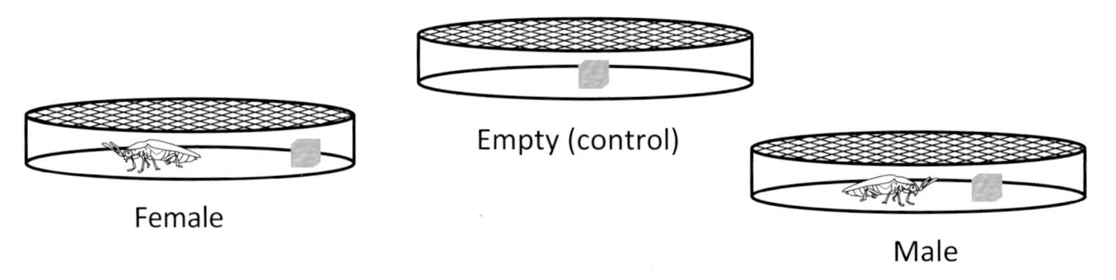


**FIGURE 11** A simple experiment that can be used to explore firefly chemical signals.

Using a similar design, you might be able to answer some additional questions:

- During different times of day, do females attract more males, and do males arrive more quickly?
- Do females show any calling behavior?
- Do males preferentially approach females from the downwind side? You can use a thread telltale to measure the wind direction, tying it to a twig or post at the same height where males are flying. Make this by attaching a small piece of Styrofoam to one end of a long piece of thread (~30 cm).
- How do males behave as they approach females? Do they fly in zigzags or in a straight line? Do they land directly on the dish or do they land nearby?
- What's the farthest distance away that males can be attracted to a female? You can mark males by gently putting a small dot on their elytra with a light-colored gel pen or an extra-fine point opaque paint marker (such as Uchida Deco-Color pens). Try releasing males at different distances, like 5, 10, and 15 meters away from the female.

## Picky Eaters?

Collecting fireflies inside a glass Mason jar is a fond childhood memory, nearly universal for those who grew up in the United States. For the most part, these Mason jar memories shine with nostalgic wonder. But often some spooky action transpires inside these jars.

At the age of five, my nephew, Nate, was traumatized by one such experience. He'd spent a delight-filled evening gathering a jarful of fireflies. At bedtime, we placed them on his nightstand so he could watch their magical glow. But when he woke the next morning, Nate was horrified to see just one large firefly remaining. The rest had vanished, leaving behind only bits of leg and wing scattered across the bottom of the jar. Frantically, he yelled for help, shouting, "My fireflies got murdered!" Like many unsuspecting firefly collectors before him, Nate had just learned firsthand about the feeding habits of *Photuris* fireflies.

It's odd enough to find an adult firefly that eats anything at all, let alone one that eats other fireflies. But *Photuris* fireflies are highly specialized predators (Figure 12). As I described in chapter 7, they've devised several hunting tactics to capture

**FIGURE 12** A *Photuris* femme fatale consumes the softer bits of a *Photinus* male (photo by Hua-Te Fang).

other fireflies, and such prey provides more than just a nutritious meal. Studies have shown that when these predatory fireflies consume *Photinus* lightningbugs, they accumulate their prey's noxious chemicals, hijacking them for protection against their own enemies (Eisner et al. 1997).

Yet we don't know very much about *Photuris* dietary predilections. Whenever they've been studied in the wild, *Photuris* fireflies seem mainly to prey on *Photinus*, occasionally on *Pyractomena*, and infrequently on other *Photuris* species (Lloyd 1984). But what do they *really* like to eat?

One June, we did an experiment to answer this question for some predatory *Photuris* active in the Great Smoky Mountains (Lewis et al. 2012). The simple setup we used consisted of several one-quart, clear plastic deli containers from the local supermarket. We poked some holes in each lid and furnished each container with a fake silk plant and some damp paper towels. Each *Photuris* female was housed in her own container. Every night for about two weeks, we offered these females assorted different insects. The menu choices included whatever fireflies happened to be active at the time—we offered several *Photinus* species, *Phausis*, and two species of *Lucidota*—along with various flies, click beetles, grasshoppers, and bugs. A reasonable compromise between natural and artificial conditions, this experiment gave prey sufficient room to find shelter and escape attack, and we could test a wider range of prey than possible in the wild.

My intention here is not to give a recap of our results—you can read our article to learn about the picky eating habits of these particular *Photuris* predators (the link is provided in the references). Instead, I hope to convince you to try this experiment at home. As you explore your local fireflies, you'll likely encounter some *Photuris*. Capture a few females and set them up in individual containers. Keep them on a natural light cycle, but out of direct sunlight. Each evening, offer them different kinds of prey and record what they do eat and what they don't. If you're curious, and not too squeamish, it's fascinating to observe these fireflies on the hunt: you can watch without disturbing them using a blue-filtered light (see gear list).

Give it a try, and let me know what you discover!

## Repatriating Fireflies

Whenever my students and I collect fireflies for scientific research, we try to replace them by repatriation. That is, we always return some of the fireflies' offspring—their eggs and/or larvae—back to their original population. If you'd like to try this approach of paying it forward, here's how.

Place a male and female together in a small (4 ounce) plastic container that you've lined with a layer of moss (for egg laying) on top of a moistened paper towel (for humidity). The container should only have a few holes; two fireflies don't require much oxygen, but they will dry out quickly if there's too much airflow. Keep them on a natural light cycle, and within a few nights they'll usually have mated (Figure 3.3 shows a pair of *Photinus* fireflies in mating position). You can check them periodically using a blue-filtered headlamp. You might also be lucky enough to find a pair outside that's already in the process of mating. If you do, don't try to pick them up. Instead, gently nudge them into your container using a small paintbrush or piece of paper, being careful not to disturb their perch. After they've mated, give them a small piece of apple, replacing this daily to prevent mold. I use organic Granny Smiths, but any apple should be fine.

Within a few days the female will deposit her tiny ivory-colored eggs, which typically measure ~1 mm across. At this point you can return everyone to its original habitat. Release the adults at the base of some plants, and find a damp location for the egg-laden moss. If you're curious to see what *Photinus* larvae look like, keep the egg-laying container in a warm, dark place. Check the eggs periodically, removing any that get moldy. In about two weeks, the eggs will hatch and tiny glowing larvae will emerge. *Photinus* larvae are difficult to raise in captivity, so you should release them back into their original habitat as soon as possible.

## Gear List: Venturing Out into the Night

Before you venture into the field to explore fireflies' nocturnal universe, it's smart to first gather some gear. Here's a firefly scientist's short list of essentials:

- A headlamp for hands-free observations of firefly behavior in the wild (I use a Petzl E89 PD Tactikka XP, with different color filters that can be swapped out). This should be equipped with a blue filter because, as explained below, fireflies' eyes are less sensitive to blue light.
- A stopwatch to measure flash timing—best if it's self-illuminating, so it won't ruin your night vision (I use a Timex Indiglo watch).
- A thermometer to measure air temperature (I got mine at my local hardware store).
- An insect net to obtain a few fireflies for identification (nets are widely available online; I use a Bioquip folding pocket net, which fits easily into a backpack).
- Some collecting containers with a bit of damp (not sopping wet) paper towel placed inside for humidity (here you could use anything, such as drugstore pill bottles).
- A small LED flashlight or penlight for talking to fireflies. The mini-LED flashlights designed to hang on keychains work well, as do the penlights that doctors use for measuring eye pupil size. It should have a switch that turns on the light only when the switch is depressed; these make it much easier to imitate firefly flash patterns.
- A small watercolor paintbrush for handling fireflies without damaging their soft bodies.
- A notebook or voice recorder for taking notes on firefly flash patterns and behavior. Dr. Andy Moiseff at the University of Connecticut wrote a free iPhone app, called Firefly Field Notes, that you can use to record flash timing, location, and weather data.
- Optional: if you're planning to watch the progression of firefly courtships, a campstool might be handy, as courtships can last a few hours.

Watching firefly behavior also requires knowing a bit about firefly vision. Fireflies don't have color vision like humans, but their eyes are tuned to be most sensitive to certain wavelengths of light. In fact, for each species, the color sensitivity of their eyes has evolved to match the color of their bioluminescent flashes (they also see ultraviolet light very well). Dusk-active fireflies, such as many *Photinus* species, mainly give off yellow flashes. So their eyes are most sensitive to yellow. On the other hand, most fireflies that become active when it's fully dark—most *Photuris*, for instance—produce green flashes, and their eyes are sensitive to green. The yellow bias of twilight-active fireflies might help them to distinguish flashes viewed against the background light reflected off green vegetation.

Observing nocturnal fireflies can be problematic—you want to be able to see them without disrupting their behavior. If you use a bright flashlight or headlamp to look closely at a firefly, it will stop behaving normally and, temporarily blinded by the light, will stop responding to flashes. The best solution is to use a blue light. Although fireflies can detect blue, their eyes are least sensitive to this color, so your light will appear very dim to them. You can convert any headlamp or flashlight into one more suitable for firefly watching simply by taping a few layers of blue acetate (or cellophane) over the lens. You can find these colored sheets at most art supply stores. By adding several layers over your light, you should be able to watch fireflies up close without disturbing them. Red headlamps are another alternative—designed to preserve human night vision, these are more widely available. But your red light will need to be very dim, because fireflies are more sensitive to red light than to blue.

Finally, a few precautions are worth mentioning. Since fireflies thrive in moist habitats, you'll often run into lots of mosquitoes too. Dress accordingly in long pants and a long-sleeved shirt. You can use insect repellent, but if it gets onto your hands, wash it off before you handle any fireflies. Avoid Buzz Off clothing—it's impregnated with the insecticide permethrin, which gets onto your skin and has neurotoxic effects on fireflies. When mosquitoes at my study site become unbearably fierce, I sometimes don a mosquito-net jacket or hat, and wear latex or nitrile protective gloves to keep them at bay. You should also be on the lookout for ticks if you're walking through tall grass or woodland edges while watching fireflies. Tuck your pant legs into your socks, and spray insect repellent on your shoes, socks, and pants. After returning home from your adventures, carefully check each layer of clothing as well as your entire body for ticks. It never hurts to ask a friend for help.

# Acknowledgments

* * *

I will always be grateful to my teachers, readers, and enablers, all of whom contributed to this book in many distinct ways. So my thanks flow out to: my teachers—Bill Bossert, for modeling intellectual enthusiasm without borders, and Peter Wayne, for keeping me on the path; my colleagues at Tufts University for building the most supportive and collegial work environment any scientist could hope for; my family in all directions for abiding my endless late-night expeditions; the light of my life, Thomas Michel, for his unflagging love and support; our two sons, Ben and Zack, who constantly remake my eyes for wonder; the town of Lincoln, Massachusetts, for preserving the natural places where fireflies thrive; my fellow firefly adventurers for sharing those nights of wonder; the students who've contributed to Tufts Team Firefly over many years; Russ Galen for galvanizing me into writing this book; Alison Kalett for her expert editorial guidance; my readers, friends, and colleagues—Jeff Fischer, Thomas Michel, John Alcock, Doug Emlen, Colin Orians, Nicole St. Clair-Knobloch, Laela Sayigh, Gwyn Loud, Karen Lewis, Francie Chew, Nooria Al-Wathiqui, Amanda Franklin, and Ben Michel—for spending their treasured time reviewing early drafts and offering priceless feedback; and the many photographers whose generosity has transformed this book into a thing of beauty. Lastly, I want to thank the citizens of the United States for supporting scientific inquiry; your tax dollars help fund our National Science Foundation, which enabled many of the discoveries described herein. Thank you all!

This book coalesced during my 2013–14 sabbatical leave from Tufts and was written on the shores of Squam Lake, New Hampshire. For this time and this place, I'm grateful.

Finally, I've long admired *Wired* magazine's colophon, the writers' monthly list of things that aided and abetted each issue. While writing this book I, too, have drawn sustenance from some tangible entities and activities: Moondog, watching phase transitions of water, critical opalescence, Corcavado National Park, TED2014, Dropbox, dragonfly metamorphosis, Pepe's white clam pizza, maitake hunting, seaglass memories, the pileated woodpecker, Eames chairs, chestnut-sided warblers, springtime forget-me-nots, Pustefix bubbles, fireworks, kayaking in the rain, and Iguaçu the Wonder Dog.

# NOTES

* * *

## CHAPTER 1: SILENT SPARKS

### A WORLD OF WONDER

To stay connected to wonder, I've drawn inspiration from the writings of Rachel Carson (Carson 1965) as well as from biologist Ursula Goodenough's compelling description of religious naturalism (Goodenough 1998).

More and more people are traveling to seek out fireflies in natural places. Information about firefly tourism comes from the following news stories. Estimates of tourist numbers in Taiwan, Thailand, and Malaysia come from e-mail correspondence with Dr. Tsung Hung Lee (Taipei; December 10, 2013), from Thancharoen (2012), and from Nada and colleagues (2009).

> Chen, R. (2012, May 19). In search of Taipei's fireflies. *Taiwan Today*. Retrieved January 15, 2015, from http://taiwantoday.tw.
>
> Brown, R. (2011, June 15). Fireflies, following their leader, become a tourist beacon. *New York Times*. Retrieved June 12, 2013, from http://www.nytimes.com/2011/06/16/us/16fireflies.html?_r=0.

The unique significance of fireflies in Japanese art, literature, and culture is described by Yuma (1993), Ohba (2004), and Oba and colleagues (2011). For translating the first of these works, I am indebted to my colleague and friend, Ray Kameda.

### FIREFLY BASICS

The estimates about when in Earth's history both beetles and fireflies originated are taken from McKenna and Farrell (2009). They're based on so-called time trees, which are evolutionary trees whose time scales have been calibrated using dated fossils.

Patterns of firefly diversity and exotic species introductions are based on Lloyd (2002, 2008), and Viviani (2001). McDermott (1964) mentions the attempted introduction of *Photuris* fireflies into parks in Seattle and Portland. Majka and MacIvor (2009) describe how European glow-worms might have gotten accidentally introduced to Nova Scotia, and report on populations they found more than fifty years later in cemeteries around Halifax.

## Looking for Love with Perfumes, Glows, and Flashes

Like all living creatures, fireflies carry their history in their genes. The evolutionary history of fireflies has been reconstructed by Branham and Wenzel (2001 and 2003) based on morphological traits, and by Stanger-Hall and colleagues (2007) based on DNA sequences.

Branham (2005) and Lewis (2009) give overviews about how different courtship styles in fireflies might have evolved.

## Further Exploration

### Silent Sparks: The TED Talk

If you're too busy to read this entire book then first watch my TED talk, which tells the condensed version of fireflies' story (14 minutes). Then come back here for a deeper dive!

https://www.ted.com/talks/sara_lewis_the_loves_and_lies_of_fireflies?language=en

### Firefly Watch

You can learn more about fireflies and sign up to report on your local lightning-bug activity with this citizen science project hosted by Boston's Museum of Science.

https://legacy.mos.org/fireflywatch/

### The Fireflyer Companion

Between 1993 and 1998, firefly expert Jim Lloyd distributed this informal newsletter dedicated to increasing awareness of firefly biology. Brimming with firefly facts, musings, poems, and even an occasional crossword puzzle, *The Fireflyer Com-*

*panion* was a good vehicle for Lloyd's eclectic and sometimes rambling communication style. You can also download these from the *Silent Sparks* blog. http://entnemdept.ufl.edu/lloyd/firefly/

## CHAPTER 2: LIFESTYLES OF THE STARS

### Deep in the Heart of the Smokies

Detailed descriptions of the life cycle, habits, and mating behavior of the Appalachian synchronous firefly *Photinus carolinus* appear in Faust (2010). Information about when, where, and how to see the Elkmont firefly display can be found on the National Park Service's website (http://www.nps.gov/grsm/learn/nature/fireflies.htm). *Photinus carolinus* is also found in Congaree National Park in South Carolina and in the Allegheny National Forest in Pennsylvania.

In his book *Sync: The Emerging Science of Spontaneous Order*, mathematician Steven Strogatz gives a highly entertaining and accessible description of the mathematical basis of synchronization and how it plays out in the engineered and natural world (Strogatz 2003).

Jon Copeland's quote is from:

Copeland, J. (1998). Synchrony in Elkmont: A story of discovery. *Tennessee Conservationist* (May–June).

The biographical material in this chapter is based on interviews I conducted with Lynn Faust in 2009, 2011, and 2013.

### Humble Beginnings

Ferris Jabr lucidly describes how insects' complicated lifestyle might have evolved and gives some historical perspective about our scientific understanding of metamorphosis.

Jabr, F. (2012, August 10). How did insect metamorphosis evolve? Scientific American online. http://www.scientificamerican.com/article/insect-metamorphosis-evolution.

Many details of the larval habits of *Lampyris noctiluca* are based on John Tyler's informative pamphlet (Tyler 2002).

## Their Glow Means No

Branham and Wenzel (2001) present phylogenetic evidence indicating that bioluminescence originated in the larval stage of some firefly progenitor, where it most likely functioned as a warning display.

## Creative Improvisation: Fireflies Evolving

I've quoted Darwin's *The Origin of Species* (1859, p. 84) for what many consider his most poetic description of natural selection.

## Synchronous Symphonies

Greenfield (2002) provides a cogent summary of various hypotheses for the evolution of synchrony in the courtship signals of various insects. Vencl and Carlson (1998) found that *Photinus pyralis* females preferentially respond to leading signals. Moiseff and Copeland (1995) looked at mechanisms of synchrony in *Photinus carolinus* fireflies, and Moiseff and Copeland (2010) showed that females responded more often to synchronous versus asynchronous male flashes.

## Further Exploration

### Darwin Online

Begun by Dr. John van Wyhe in 2002, this site provides digital, searchable versions of Charles Darwin's books, field notes, journals, and more, along with downloadable audio and image files.
http://darwin-online.org.uk/

### Darwin in Print

Wilson, E. O., editor (2006). *From So Simple a Beginning: Darwin's Four Great Books*. W. W. Norton, New York, NY. 1706 pp.

Eminent biologist and Pulitzer Prize–winning author E. O. Wilson has gathered together and annotated four of Darwin's works in a beautifully illustrated and affordable volume: *Voyage of the H.M.S. Beagle* (1845), *The Origin of Species* (1859), *The Descent of Man and Selection in Relation to Sex* (1871), and *The Expression of Emotions in Man and Animals* (1872). In his afterword, Wilson thoughtfully examines divisions between science and religious belief.

### The Life and Times of European Glow-Worms

The lifestyle of *Lampyris noctiluca* is showcased in vivid detail in two books by talented naturalists. The second is among the last works written by the great French entomologist J. Henri Fabre; though its literary style is rather florid for modern tastes, it remains entertaining.

John Tyler (2002). *The Glow-worm*. Privately published.

Fabre, J. H. (1924) *The Glow-worm and Other Beetles*. Dodd, Mead, New York, NY.

### The UK Glow Worm Survey

This informal group was set up Robin Scagell in 1990 to gather information about glow-worm sightings throughout the United Kingdom. The website describes the biology and conservation of these glow-worms and provides links to many other resources and books.
http://www.glowworms.org.uk/

### *Earth-Born Stars: Britain's Secret Glow-Worms*

This evocative short film by Christopher Gent shows larvae feeding on snails, illustrates female courtship habits, and explores conservation threats to one of Britain's most adored yet mysterious insects.
https://vimeo.com/31952006

## CHAPTER 3: SPLENDORS IN THE GRASS

### Wild about Fireflies

*Photinus* fireflies were the main subject of Jim Lloyd's doctoral thesis (Lloyd 1966), where he described their geographic and habitat distributions, courtship behavior, and more. The frontispiece from this work is shown in Figure 3.1 (used with permission from University of Michigan's Museum of Zoology), which artfully illustrates the male flight paths and flash patterns for the following *Photinus* species: (1) *consimilis* slow-pulse (2) *brimleyi*, (3) *consimilis* fast-pulse, (5) *marginellus*, (6) *consanguineus*, (7) *ignitus*, (8) *pyralis*, and (9) *granulatus*.

### Defining the Indefinable

Charles Darwin's quote is from a letter written to his good friend and confidant, the botanist Joseph Hooker.

Darwin, C. R., Letter to J. D. Hooker. December 24, 1856. *Darwin Correspondence Database*. http://www.darwinproject.ac.uk/entry-2022.

### Heading Out into the Night

Carl Zimmer highlighted our firefly research in his award-winning article in the *New York Times*:

Zimmer, C. (2009, 29 June). Blink twice if you like me. *New York Times*. Retrieved from http://www.nytimes.com/2009/06/30/science/30firefly.html.

We describe the courtship behaviors of *Photinus greeni* fireflies in Demary and colleagues (2006) and Michaelidis and colleagues (2006).

### A Light Snack

Lloyd (2000) reports tracking a couple hundred firefly males belonging to *Photinus collustrans* to see how likely they were to find a female versus encounter a predator. Various predators that feed on fireflies are described in Lloyd (1973a), Day (2011), and Lewis and colleagues (2012).

### Closer Encounters

In Lewis and Wang (1991), we delve into the courtship and mating behavior of two New England fireflies, *Photinus marginellus* and *Photinus aquilonius*.

### To the Victors Go the Spoils

Trivers (1972) suggested that differences in male and female sexual behavior resulted from the asymmetry in parental investment between the sexes. Biologist Darryl Gwynne and his colleague won an IgNobel Prize ("Achievements that make people laugh, and then think") for discovering some males that aren't choosy at all; in the Australian beetle *Julodimorpha bakervelli*, males often copulate with discarded beer bottles along the roadside (Gwynne and Rentz 1983).

Erica Deinert kindly showed me mate-guarding *Heliconius* butterflies in Costa Rica. Lynn Faust has described male pupal-guarding behavior in two fireflies, *Photinus carolinus* (Faust 2010), and *Pyractomena borealis* (Faust 2012).

Courtship is fiercely competitive for many *Photinus* fireflies, as described by Maurer (1968), Vencl and Carlson (1998), and Faust (2010). The hooked wing covers of male *Pteroptyx* fireflies, which clamp around the female's abdomen during mating, were described by Wing and colleagues (1982). Lloyd (1979a) described the pseudo-female flash responses sometimes given by firefly males that have been unsuccessful in finding a female.

## Ladies' Choice

The Darwin quote is from Part II, p. 38, of his book, *The Descent of Man and Selection in Relation to Sex* (1871), in which he describes what sexual selection is all about.

Fisher (1930) first modeled how female choice could trigger the elaboration of extravagant male bits such as the peacock's tail. Female choice in *Photinus* fireflies has been demonstrated using photic playback experiments conducted by Branham and Greenfield (1996), Cratsley and Lewis (2003), and Michaelidis and colleagues (2006). Demary and colleagues (2006) show that females preferentially give response flashes to certain males, and these males subsequently achieve higher mating success. Lewis and Cratsley (2008) provide a somewhat technical review of what scientists have learned about flash signal evolution, courtship, and predation in fireflies.

## Trading Places

Lewis and Wang (1991) describe the seasonal shift in firefly sex ratios, and Cratsley and Lewis (2005) show that late-season males generally choose females who have more eggs.

## Further Exploration

### More on Sexual Selection

In the second part of his 1871 book, Darwin describes the power of sexual selection to shape animal form and function. After outlining the principles and mechanisms of this fascinating evolutionary process, in separate chapters Darwin describes how sexual selection has led to the evolution of many wonderful and sometimes bizarre male features in crustaceans, mollusks, insects, amphibians, reptiles, birds, and even humans.

Darwin, C. (1871). *The Descent of Man and Selection in Relation to Sex*. John Murray, London.

http://darwin-online.org.uk/converted/pdf/1871_Descent_F937.2.pdf

Here are two more excellent accounts of sexual selection—both shorter and also a bit wittier than Darwin's. Under the guise of a sex advice column for lovelorn beetles, stick insects, stalk-eyed flies, mice, and manatees, Olivia Judson's hilarious book describes some of the weird structures and behaviors that have evolved as a result of sexual selection.

Judson, O. (2002). *Dr. Tatiana's Sex Advice to All Creation*. Metropolitan Books, Henry Holt, New York, NY. 320 pp.

Cronin, H. (1993). *The Ant and the Peacock*. Cambridge University Press, New York, NY. 504 pp.

### Traveling the Firefly Trail with Jim Lloyd

Jim Lloyd's monograph on US *Photinus* fireflies describes their geographic and habitat distributions, courtship flash behavior, and more, and is available free online:

Lloyd, J. E. (1966). Studies on the flash communication system in *Photinus* fireflies. *University of Michigan Miscellaneous Publications* 130: 1–95.

http://deepblue.lib.umich.edu/handle/2027.42/56374.

Between 1997 and 2003, Jim Lloyd published a series of articles in the open-access scientific journal *Florida Entomologist*, called "On Research and Entomological Education." Written in the form of letters addressed to his students, these ramblings are packed with information about firefly natural history, and are chock-full of ideas for field studies.

Lloyd, J. E. (1997). On research and entomological education, and a different light in the lives of fireflies (Coleoptera: Lampyridae; *Pyractomena*). *Florida Entomologist* 80: 120–31. http://journals.fcla.edu/flaent/article/view/74752.

Lloyd, J. E. (1998). On research and entomological education II: A conditional mating strategy and resource-sustained lek(?) in a classroom firefly (Coleoptera: Lampyridae; *Photinus*). *Florida Entomologist* 81: 261–72. http://journals.fcla.edu/flaent/article/view/74829.

Lloyd, J. E. (1999). On research and entomological education III: Firefly brachyptery and wing "polymorphism" at Pitkin marsh and watery retreats near summer camps (Coleoptera: Lampyridae; *Pyropyga*). *Florida Entomologist* 82: 165–79. http://journals.fcla.edu/flaent/article/view/74877.

Lloyd, J. E. (2000). On research and entomological education IV: Quantifying mate search in a perfect insect-seeking true facts and insight (Coleoptera: Lampyridae, *Photinus*). *Florida Entomologist* 83: 211–28. http://journals.fcla.edu/flaent/article/view/59545.

Lloyd, J. E. (2001). On research and entomological education V: A species (c)oncept for fireflyers, at the bench and in old fields, and back to the Wisconsian glacier. *Florida Entomologist* 84: 587–601. http://journals.fcla.edu/flaent/article/view/75006.

Lloyd, J. E. (2003). On research and entomological education VI: Firefly species and lists, old and now. *Florida Entomologist* 6: 99–113. http://journals.fcla.edu/flaent/article/view/75180.

## CHAPTER 4: WITH THIS BLING, I THEE WED

### After the Lights Go Out

The demise of monogamy was reported by Natalie Angier in the *New York Times* in 1990. Some scientific ramifications of the "Polyandry Revolution" are explored in a theme issue introduced by Pizzari and Wedell (2013). We describe polyandry among *Photinus* fireflies in the wild in Lewis and Wang (1991).

Angier, N. (1990, August 21). Mating for life? It's not for the birds or the bees. *New York Times*. http://www.nytimes.com/1990/08/21/science/mating-for-life-it-s-not-for-the-birds-of-the-bees.html.

### Sperm Wars, Sperm Love

Leigh Simmons, an accomplished evolutionary biologist, provides a comprehensive review of sperm competition theory and mechanisms in Simmons (2001).

Quote about male genitalia and Swiss army knives is from p. 22 of Lloyd (1979b). Waage (1979) describes sperm removal in the damselfly *Calopteryx maculata*, and Davies (1983) reports cloacal pecking and sperm ejection by dunnocks, *Prunella modularis*.

Evidence for sexual selection via cryptic female choice is presented in two books, Eberhard (1996) and Peretti and Aisenberg (2015).

## Amorous Bundles

We reported our discovery of nuptial gifts for *Photinus* fireflies in van der Reijden and colleagues (1997), and later for Japanese fireflies in South and colleagues (2008).

## Finding the Perfect Gift

Different kinds of nuptial gifts, along with their evolutionary causes and consequences, are described in Lewis and South (2012) and Lewis and colleagues (2014). Albo and colleagues (2011) report that in the spider *Pisaura mirabilis*, males offering gift-wrapped worthless (inedible) gifts were as successful in obtaining mates as males that offered genuine gifts (dead flies). But females terminate matings sooner when males deceive them with worthless gifts, so these males have a disadvantage in sperm competition. Chemical ecologists Tom Eisner and Jerry Meinwald (1995) relate the astounding story of how ornate moths, *Utetheisa ornatrix*, sequester bitter-tasting alkaloids from their larval food plants, and how males then pass these toxins along to females in their nuptial gifts. Some intimate events that transpire during snail sex have been elucidated by Koene (2006).

## Male Sexual Economics

We tested the notion that producing nuptial gifts is costly for male fireflies in Cratsley and colleagues (2003). In South and Lewis (2012), we found that males gain a paternity benefit from giving larger gifts because they get to sire a greater percentage of offspring.

## Bright Lights and Bling: What's in It for the Female?

Demonstrating how valuable a nuptial gift can be when nutrients are scarce, Yoshizawa and colleagues (2014) describe cave-dwelling Brazilian bark lice whose female intromittent organ appears to have evolved through competition for male nuptial gifts.

As discussed in Lewis and colleagues (2004), gift giving plays an important role in firefly economics because most fireflies stop eating once they've become adults. In Rooney and Lewis (1999), we describe our radio-labeling studies showing that a female uses protein from the male's gift to help provision her eggs. In Rooney and Lewis (2002), we show that *Photinus ignitus* females who mate more often subsequently lay more eggs. In *Photinus greeni* females, a different benefit showed up: females who received larger gifts lived longer (South and Lewis 2012).

Cratsley and Lewis (2003) discovered a correlation between males' flash duration and their spermatophore size in *Photinus ignitus* fireflies; these females might be able to use a male's flash signals to predict what size gift he could offer. But when we looked for a similar relationship in *Photinus greeni* fireflies, we found none (Michaelidis et al. 2006).

## Further Exploration

### Animal Nuptial Gifts

In this short article we describe the astonishing diversity of nuptial gifts found across the animal kingdom:

> Lewis, S. M., A. South, N. Al-Wathiqui, and R. Burns (2011). Quick guide: Nuptial gifts. *Current Biology* 21: 644–45. http://www.sciencedirect.com/science/article/pii/S096098221100604X.

This excellent article published on Valentine's Day by Brandon Keim in *Wired* online goes beyond the ordinary to illustrate animal nuptial gifts:

> Keim, B. (2013, February 14). Freaky ways animals woo mates with gifts. *Wired*. http://www.wired.com/2013/02/valentines-day-animal-style/.

From Dutch biologist and science writer Menno Schilthuizen come two informative books, both written with enthusiasm, clarity, and humor. *Nature's Nether Regions* describes the reproductive equipment of barnacles, slugs, apes, and more, explaining how the bizarre and unusual inventions known as animal genitalia have been forged under postcopulatory sexual selection. His earlier book, *Frogs, Flies, and Dandelions* describes the history of ideas about how new species emerge, and explores the role that sexual selection might play in the speciation process.

Schilthuizen, M. (2014). *Nature's Nether Regions*. Viking, New York, NY. 256 pp.

Schilthuizen, M. (2001). *Frogs, Flies, and Dandelions: Speciation—The Evolution of New Species*. Oxford University Press, Oxford. 256 pp.

## CHAPTER 5: DREAMS OF FLYING

### Into the Umwelt

The physiologist Jakob von Uexküll wrote a charming treatise exploring the sensory-perceptual worlds of different animals, as he explained and illustrated his concept of Umwelt (von Uexküll 1934). Healya and colleagues (2013) suggest that an organism's perception of time depends on its size and metabolic rate, and find support for this idea using a phylogenetic comparison across different vertebrates.

### Sexual Dimorphism

A deep exploration of anglerfish sex appears in an article by a world expert on such fish, Ted Pietsch (2005). South and colleagues (2011) report our discovery that in fireflies, females' flight ability is correlated with male nuptial gifts.

### The King of Glow

The cultural association of Saint John's Eve with glow-worms is described in Raphaël De Cock's review of the biology and behavior of European fireflies (De Cock 2009). The biographical material in this chapter is based on interviews with Raphaël De Cock that I conducted in 2011 and 2013.

Raphaël De Cock's studies on luminescence acting as a warning signal for toads are reported in De Cock and Matthysen (2003). De Cock and Matthysen (2005) present evidence that females in the lesser European glow-worms use pheromones to attract mates.

### Ghostly Glows and Phantom Fumes

Frick-Ruppert and Rosen (2008) describe blue ghost natural history and behavior. The fairies quote is taken from an entry on the blog called *Blue Ghost Post*, about Bennie Lee Sinclair and Don Lewis's Firefly Forest in South Carolina.

Blueghoster. *Saints, Sanctuaries, and The Blue Ghosts*. April 6, 2010. http://blueghostpost.blogspot.com/2010/04/saints-santuaries-and-blue-ghosts.html.

The scientific report describing our blue ghost studies appears in De Cock and colleagues (2014). Dr. Somyot Silalom told me about the egg-guarding behavior of *Lamprigera tenebrosus* females. Studies on *Rhagophthalmus ohbai*, regarded as close kin to fireflies, reported that these bioluminescent, egg-guarding females release a volatile chemical that protects their eggs against attack by soil microbes (Hosoe et al. 2014).

## FURTHER EXPLORATION

### THE GLOW-WORM SONG

Here you can enjoy "Glow Little Glow-Worm" as recorded by the Mills Brothers in 1952.
http://www.youtube.com/watch?v=2zOoAPn3OjQ

### SEXUAL DIMORPHISM

This authoritative book by Daphne Fairbairn, an evolutionary biologist at the University of California, Riverside, explores the causes and consequences of the vast size differences between the sexes in particular species, including elephant seals, garden spiders, barnacles, and anglerfish.

Fairbairn, D. J. (2013). *Odd Couples: Extraordinary Differences between the Sexes in the Animal Kingdom*. Princeton University Press, Princeton, NJ. 312 pages.

Matt Simon describes anglerfish sexual habits in an article profusely illustrated with photographs (and a video) of dead fish.

Simon, M. (2013, November 8). Absurd creature of the week: the anglerfish and the absolute worst sex on Earth. *Wired*. http://www.wired.com/2013/11/absurd-creature-of-the-week-anglerfish/.

### DUPONT STATE FOREST

The history of DuPont State Forest in North Carolina, located between Hendersonville and Brevard (http://www.dupontforest.com/) is described in this article. Formerly the site of a DuPont manufacturing facility, this state forest encompasses 4,200 hectares of preserved lands and is a good spot to see blue ghost fireflies during May.

Summerville, D. (2011). Southern Lights: Blue Ghost Fireflies. *Our State: North Carolina*. http://www.ourstate.com/lightning-bugs/.

### More on Blue Ghost Fireflies

These two reports describe what we currently know about blue ghost fireflies. Our 2014 article includes a supplementary video of blue ghost mating behavior (http://journals.fcla.edu/flaent/article/view/83837).

Frick-Ruppert, J., and J. Rosen (2008). Morphology and behavior of *Phausis reticulata* (Blue Ghost Firefly). *Journal of North Carolina Academy of Science* 124: 139–47. http://dc.lib.unc.edu/cdm/ref/collection/jncas/id/3883.

De Cock, R., L. Faust, and S. M. Lewis (2014). Courtship and mating in *Phausis reticulata* (Coleoptera: Lampyridae): Male flight behaviors, female glow displays, and male attraction to light traps. *Florida Entomologist* 97: 1290–307. http://dx.doi.org/10.1653/024.097.0404.

## CHAPTER 6: THE MAKING OF A FLASHER

### A Chemistry Set for Light

Reviews by Wilson and Hastings (1998, 2013) outline the chemistry of bioluminescence. The rendition of luciferase is from David S. Goodsell's June 2006 Molecule of the Month in the RCSB's Protein Data Bank (http://dx.doi.org/doi:10.2210/rcsb_pdb/mom_2006_6). Niwa and colleagues (2010) measured quantum yields of 40–60% from firefly bioluminescence.

### Firefly Lights Evolving

Viviani (2002) and Oba (2015) review ideas about how beetle luciferases might have evolved. Yuichi Oba and his colleagues (2008) measured luciferin content of various nonluminescent beetles. Lynch (2007) describes the role of gene duplication in the evolution of snake venom. Darwin's prescient quote about exaptation is from the *Origin of Species* (1859, p. 190).

### Putting Fireflies to Work

Practical applications for fireflies' light-producing ability are described in Weiss (1994), Rosellini (2012) and Andreu and colleagues (2013).

Weiss, R. (1994, August 29). Researchers gaze into the (insect) light and gain answers. *Washington Post*, A3.

### Controlling the Flash

John Buck's biographical information is from Case and Hanson (2004), and from his *New York Times* obituary.

Pearce, J. (2005, April 3). John B. Buck, who studied fireflies' glow, is dead at 92. *New York Times*. http://www.nytimes.com/2005/04/03/science/03buck.html.

### A Journey inside the Firefly Lantern

Buck (1948) and Ghiradella (1998) provide a detailed look at the internal anatomy of the firefly lantern. Not only an expert on firefly lantern anatomy, Helen Ghiradella is also a skilled artist whose drawings allow others to appreciate the internal architecture of this light-producing marvel. Figure 6.2 is modified with permission from Ghiradella (1998).

### Discovering the Firefly's Light Switch

Our discoveries about the role of nitric oxide in firefly flash control were reported in Trimmer et al. (2001). A complementary hypothesis about oxygen control based on the anatomy of firefly air tubes is presented by Ghiradella and Schmidt (2008).

### Getting in Sync

Smith (1935) describes witnessing the synchronous fireflies (*Pteroptyx*) in Thailand, although he incorrectly concluded this activity was unrelated to mating.

John and Elisabeth Buck's trip to Thailand produced the very first scientific study of how *Pteroptyx malaccae* fireflies manage to flash synchronously (Buck and Buck 1968). Elisabeth Buck's quote is from the Radiolab podcast, *Emergence*, broadcast on February 18, 2005. Bookending fifty years of scientific research on the mechanisms of firefly flash synchrony, John Buck wrote two review articles about the different physiological mechanisms underlying flash synchrony: Buck (1938) and Buck (1988). In the latter, he also describes the geographic and taxonomic distribution of several types of flash synchrony.

## Science Confidential

Niko Tinbergen (1907–1988) was a Dutch ethologist and ornithologist who shared the 1973 Nobel Prize in Physiology or Medicine for his discoveries about animal behavior. His 1963 paper was dedicated to Konrad Lorenz on the occasion of the latter's sixtieth birthday. My account of this scientific feud is based on personal conversations and correspondence exchanged with both Jim Lloyd and John Buck. Buck and Buck (1978) focus their attention on how synchrony might benefit a group of males, while Lloyd (1973b) focuses attention on what advantages synchrony might provide to individual males as well as to the group. Faust (2010) describes how *Photinus carolinus* males switch from flashing synchronously to flashing chaotically as they approach a female. Case (1980) describes some close-up behavioral interactions inside *Pteroptyx* display trees.

## Further Exploration

### More about Bioluminescence

Living creatures that make their own light are deeply fascinating. The 2009 science fiction movie *Avatar*, written and directed by James Cameron, is well known for the exuberantly bioluminescent creatures living on the fantasy world called Pandora. Two leading authorities on bioluminescence, Thérèse Wilson and Woody Hastings, describe the molecular details of bioluminescence in selected creatures, including fireflies.

Wilson, T., and J. W. Hastings (2013). *Bioluminescence: Living Lights, Lights for Living*. Harvard University Press, Cambridge, MA. 208 pp.

The *Bioluminescence Web Page* (http://www.lifesci.ucsb.edu/~biolum/) celebrates all living things that produce their own light (hosted by the University of California at Santa Barbara).

### More about Synchrony

Writing for a general audience, leading mathematician and an award-winning science communicator Steve Strogatz explains how thousands of fireflies, cardiac pacemaker cells, or electrons in a superconductor manage achieve their highly organized synchronous behavior without a conductor.

Strogatz, S. H. (2003). *Sync: The Emerging Science of Spontaneous Order*. Hyperion Books, New York, NY. 338 pp.

Biologist Michael Greenfield takes a detailed look at the acoustic, chemical, vibratory, visual, and bioluminescent signals that insects use to communicate with one another. He also reviews the hows and whys of collective male synchrony in fireflies, crickets, and cicadas.

Greenfield, M. (2002). *Signalers and Receivers: Mechanisms and Evolution of Arthropod Communication*. Oxford University Press, New York, NY. 432 pp.

### Radiolab: Emergence

This February 18, 2005 Radiolab podcast explores how individuals following simple rules can generate complex group behaviors, like firefly synchrony. This episode features interviews with biologists John and Elisabeth Buck, as well as with mathematician Steve Strogatz.
http://www.radiolab.org/story/91500-emergence/

## CHAPTER 7: POISONOUS ATTRACTIONS

### For the Love of Insects

Most of Tom Eisner's quotations were taken from video interviews he gave in 2000 for the *Web of Stories* project (see Further Explorations), or from this 2003 NPR interview. Others were from personal conversations I had with Tom during a visit to Cornell in 2008.

Eisner, T. (2003). Interviewed by Robert Siegel on *All Things Considered*, National Public Radio, November 18, 2003. http://www.npr.org/templates/story/story.php?storyId=1511501.

## Lightningbugs for Breakfast? No-No!

In addition to the *Web of Stories* video, Eisner relates a charming version of the Phogel story in his popular science book, *For Love of Insects* (Eisner 2003).

Jim Lloyd amassed more than a century's worth of anecdotal evidence about which creatures do and do not eat fireflies (Lloyd 1973). Knight and colleagues (1999) give detailed case histories about bearded dragons who met gruesome deaths by firefly. The bat study (Moosman et al. 2009) was conducted prior to the 2007 outbreak of white nose syndrome, a disease that has devastated bat populations in the eastern United States.

## Chemical Weapons

The chemical defenses called lucibufagins were first identified by Tom Eisner and his colleagues (1978) from adults of three *Photinus* fireflies, which make these steroidal pyrones in several flavors. Lucibufagins have also been reported in adults of the diurnal firefly, *Lucidota atra* (Gronquist et al. 2006) as well as in larvae of *Lampyris noctiluca* (Tyler et al. 2008). Day (2011) provides a review of firefly defenses.

Gao and colleagues (2011) provide an overview of the therapeutic potential of bufadienolide drugs, and Banuls et al. (2013) discuss the antitumor activity of thirty-five such compounds.

## A Multifaceted Defense Strategy

Reflex bleeding was first described in *Photinus pyralis* by Blum and Sannasi (1974), and the phenomenon has since been reported from several other firefly genera, including *Pyrocoelia, Luciola*, and *Lucidina*. The pop-out defensive glands of firefly larvae were described for *Luciola leii* by Xinhua Fu and his colleagues (2007), and then again for several additional species in Fu et al. (2009). The dark side of the Smokies Light Show is explored in Lewis et al. (2012).

## Evolution of Warning Displays

The many years that Alfred Russel Wallace spent doing fieldwork in the tropics helped him grasp the concept of warning coloration more readily than did Darwin. He described cryptic coloration, warning displays, and mimicry in his 1867 article (quote appears on p. 9). Wallace used the term "danger-flag" in his 1889 book on natural selection (p. 232). The Darwin quote is from a letter written at Down House to A. R. Wallace dated February 26, 1867 (F. Darwin 1887, p. 94)

De Cock and Matthysen (2001) showed that firefly color patterns act as a warning signal for starlings. Other studies have demonstrated the power of glows to facilitate avoidance learning in toads (De Cock and Matthysen 2003), mice (Underwood et al. 1997), spiders (Long et al. 2012), and bats (Moosman et al. 2009).

## Firefly Look-Alikes: Tasty or Toxic?

Bates wrote extensively about his travels and natural history observations (see Further Explorations). Quotation is from his 1862 paper on mimicry among Amazonian butterflies (Bates 1862, 507).

The photographs of firefly mimics in Figure 7.3 are (clockwise from upper left): a cockroach (Blattellinae), *Pseudomops septentrionalis* (photo by John Hartgerink); a blister beetle (Meloidae), *Pseudozonitis* sp. (photo by Mike Quinn, TexasEnto.net); a longhorn beetle (Cerambycidae), *Hemierana marginata* (photo by Patrick Coin), a net-winged beetle (Lycidae), *Plateros* sp. (photo by Gayle and Jeanell Stickland), a moth (photo by Shirley Sekarajasingham), and a soldier beetle (Cantharidae), *Rhagonycha lineola* (photo by Patrick Coin).

## The Vampire Firefly

Aggressive mimicry by *Photuris* femmes fatales was described by Lloyd (1965, 1975, 1984). Tom Eisner and his colleagues (1997) showed that *Photuris* females sequester their prey's toxic lucibufagins, stockpiling them for their own self-protection. While femmes fatales acquire the vast majority of their lucibufagin from prey, these researchers also found tiny amounts of lucibufagin in some lab-reared *Photuris* that never had access to *Photinus* prey. Andres González and his colleagues (1999) noted that *Photuris* larvae have an endogenous defensive chemical, known as betaine, which carries over into adults and affords them some limited protection against predators. They also showed that females endow their eggs with high concentrations of the lucibufagins that they have sequestered from their prey.

Lloyd and Wing (1983) and Woods and colleagues (2007) describe hawking by *Photuris* predators, and Faust and colleagues (2012) describe the kleptoparasitic behavior of these thieves in the night.

## Further Exploration

### More about Tom Eisner

Web of Stories

This online repository contains video interviews with some of the greatest scientists of our time. Tom Eisner is one, and he talks about his life and work in several segments. In *Why Entomologists Eat Bugs: A Firefly Story*, Eisner tells how he and Phogel discovered the chemicals that help fireflies defend themselves against predators.
http://www.webofstories.com/play/thomas.eisner/7

A gifted communicator, Eisner's 2003 book provides an entertaining account of his many explorations in the land of chemical ecology, vividly illustrated with his own spectacular photographs.

Eisner, T. (2003). *For Love of Insects*. Belknap Press of Harvard University Press, Cambridge, MA. 464 pp.

### Light Snacks: Predation on Fireflies

This short video by the Lewis Lab shows field observations of fireflies under attack by spiders, bugs, and the predatory *Photuris* femmes fatales, based on research by Sara Lewis, Lynn Faust, and Raphäel De Cock. The Griff Sextet provided the soundtrack, which features scientist and musician Raphäel De Cock on vocals.
http://vimeo.com/28816083

### More about Warnings and Mimicry

Ruxton and colleagues provide an outstanding and readable explanation, though perhaps a bit technical, of what's known concerning the evolution of cryptic coloration, warning signals, and mimicry in animals.

Ruxton, G. D., T. N. Sherratt, and M. P. Speed (2004). *Avoiding Attack: The Evolutionary Ecology of Crypsis, Warning Signals, and Mimicry.* Oxford University Press, Oxford.

Quite popular when first published in 1863, this nineteenth-century classic chronicles travels through the Amazon basin by Henry William Bates, the British naturalist and insect collector. His wide-ranging and charming account covers natural history, geography, ethnography, and more. One admirer, Charles Darwin, called it "the best work of natural history travels ever published in England." A hundred fifty years down the road, it remains an enviable model of lyrical nature writing.

Bates, H. W. (2009). *The Naturalist on the River Amazon*. Cambridge University Press, Cambridge, UK.

## CHAPTER 8: LIGHTS OUT FOR FIREFLIES?

### Darkening Summers

Keneagy (1993) reported on dwindling numbers of fireflies in Florida. The firefly e-mails appeared on a website maintained by Donald Ray Burger, a personal injury lawyer and firefly fan based in Houston, Texas. This site (http://www.burger.com/firefly.htm) provides links to lots of useful firefly information; since 1996 Burger has been collecting and posting hundreds of reports he receives about firefly numbers from all over North America.

Keneagy, B. (1993, September 25). Lights out for firefly population. *Orlando Sentinel*. http://articles.orlandosentinel.com/1993-09-25/news/9309250716_1_lightning-bugs-fireflies-osceola.

Estimates of firefly declines in Thailand are from Casey (2008) and New Tang Dynasty TV's news story about declining firefly populations along the Mae Klong River in Bam Lomtuan, southern Thailand.

Casey, M. (2008, August 30) Lights out? Experts fear fireflies are dwindling. *USA Today*. http://usatoday30.usatoday.com/news/world/2008-08-30-1331112362_x.htm.

New Tang Dynasty Television (2009, June 10). *Fireflies' spectacle coming to an end*. (Video file). http://www.youtube.com/watch?v=06RHumVQ-e8.

### Paved Paradise

Jim Lloyd's quote about the absence of fireflies in Houston appeared in a news story by Grossman (2000).

Grossman, W. (2000, March 2). Fireflies are disappearing from the night sky. *Houston Press*. Retrieved from http://www.houstonpress.com/issues/2000-03-02/feature2.html.

Jusoh and Hashim (2012) describe how the loss of mangrove habitat has affected the Malaysian synchronous firefly, *Pteroptyx tener*, and Thancharoen (2012) discusses firefly tourism and conservation in Thailand.

Sonny Wong provides advice about good firefly-watching behavior on his blog about Malaysian fireflies: https://malaysianfireflies.wordpress.com/2010/01/20/firefly-watching-ethics/. His quote is from an interview with Sharmilla Ganesan (2010).

> Ganesan, S. (2010, February 16). Keeping the lights on. *The Star Online*. http://www.thestar.com.my/Lifestyle/Features/2010/02/16/Keeping-the-lights-on/.

### Drowning in Light

David Owen (2007) wrote about the International Dark-Sky Association.

> Owen, D. (2007, August 20). The dark side: Making war on light pollution. *New Yorker*. http://www.newyorker.com/magazine/2007/08/20/the-dark-side-2.

Ineichen and Rüttimann (2012) describe how artificial light affects European glow-worms. Rich and Longcore (2006) explore the ecological consequences of artificial illumination, including effects on nest choice and breeding success of birds to behavioral and physiological changes in salamanders. One chapter by Jim Lloyd speculates about how stray light could affect fireflies.

### A Bounty on Firefly Lights

Pieribone and Gruber (2005, p. 101) reproduced a photo taken at Johns Hopkins "before the molecular biology revolution" with William McElroy sitting next to an enormous pile of bounty-hunted fireflies, preparing to extract their luciferase. I've met some people who collected for McElroy when they were kids, and they still recall the excitement of running around Baltimore at night, gathering fireflies to trade in for cash the next day.

The *Chicago Tribune* reported on the Sigma Chemical Company's firefly-collecting activities in 1987, as did Valerie Reitman in the *Wall Street Journal* in 1993. Searching for "Firefly Scientist's Club" in Google News will turn up many old newspaper advertisements soliciting collectors to help in harvesting fireflies for commercial sale.

United Press International (1987, August 24). Pennies from heaven for firefly catchers. *Chicago Tribune*. http://articles.chicagotribune.com/1987-08-24/business/8703040337_1_fireflies-sclerosis-and-heart-disease-shark-tank.

Reitman, V. (1993, September 2) Scientists are abuzz over the decline of the gentle firefly. *Wall Street Journal*. A1.

As of 2015, many firefly-derived products are still listed on the Sigma-Aldrich website, including:

Dried firefly tails (abdomens) http://www.sigmaaldrich.com/catalog/product/sigma/fft.

Whole dried fireflies http://www.sigmaaldrich.com/catalog/product/sigma/ffw.

Gilbert (2003) describes firefly-collecting efforts in Morgan County, Tennessee, led by Pastor Dwight Sullivan from Whittier, California. As reported by O'Daniel (2014), during the summer of 2014 Sullivan was paying $2 per 100 live fireflies collected. In Bauer et al. (2013), we developed a model to predict whether firefly populations could sustain various levels of harvesting.

Gilbert, K. (2003, June 20). Fireflies light the way for this pastor. *United Methodist Church News*. http://archives.umc.org/umns/news_archive2003.asp?ptid=&story={661B5CCE-59B8-4C1F-8BF3-F0F17B99DDE6}&mid=2406.

O'Daniel, R. (2014, July 16). Blicking bucks: Scientists will pay for summer's glow. *Morgan County News*. http://www.morgancountynews.net/content/blinking-bucks-scientists-will-pay-summers-glow.

## Other Insults

Estimates of pesticide application rates are from Beyond Pesticide's website, which provides fact sheets, news, and advocacy around the human health and environmental effects of pesticides:

Beyond Pesticides Fact Sheet (August 2005). *Lawn Pesticides Facts and Figures*. http://beyondpesticides.org/lawn/factsheets/LAWNFACTS&FIGURES_8_05.pdf.

Ki-Yeol Lee and colleagues (2008) provide a comprehensive study that experimentally measured how pesticides and fertilizers affect the different life stages of the common Asian firefly, *Luciola lateralis* (now renamed *Aquatica lateralis*); this firefly spends its larval stage living underwater. Analyzing the time course of firefly declines, Masahide Yuma (1993) pointed to increasing pesticide use on rice fields as a factor contributing to declining populations of Japanese fireflies. I am indebted to my former student and now colleague Ray Kameda for translating this and other material from Japanese.

## Hotaru Koi: Come Firefly!

Erik Laurent (2001) and Akito Kawahara (2007) provide excellent descriptions of Japanese entomophilia. The deep appreciation felt for fireflies in Japanese culture is described by Yuma (1993), Ohba (2004), and Oba and colleagues (2011).

Lafcadio Hearn (1850–1904) was a popular writer, translator, and interpreter of Japanese life and culture. The quotes are from his 1902 piece "Fireflies" reprinted on pp. 188–94 in Allen and Wilson's 1992 anthology of Hearn's writings.

Yuma (1995) traces the history of Uji's fireflies. The late-night oviposition behavior of Genji-botaru females was described by Yuma and Hori (1981). The Tokyo Hotaru Festival (http://tokyo-hotaru.jp/) is now held annually in May.

> Spacey, J. (2012, June 14). Hotaru Festival: A light spectacular in Tokyo. *Japan Talk*. http://www.japan-talk.com/jt/new/tokyo-hotaru-festival.

Iguchi (2009) reports evidence that local fireflies in the town of Tatsuno, Nagano Prefecture, which holds an annual summer firefly festival, have been contaminated by releasing nonnative, artificially bred Genji fireflies from other regions.

## Further Exploration

### About Fireflies in Japan

*Beetle Queen Conquers Tokyo* is a fascinating 2009 documentary produced and directed by Jessica Oreck, which takes a closer look at the Japanese enthusiasm for insects, especially beetles.

The Tokyo Hotaru Festival is featured in this short video (https://vimeo.com/67980309), which shows artificial fireflies being launched onto the Sumida River as it flows through downtown.

### Selangor Declaration on the Conservation of Fireflies

Written by an international group of firefly experts in 2010 in Selangor, Malaysia, this declaration was updated in 2014.
http://www.lampyridjournal.com/the-selangor-declaration-conservation-of-fireflies/

### The International Dark-Sky Association

This nonprofit group works to spread the word about light pollution and provides resources to help preserve the night.
http://www.darksky.org/

## FIELD GUIDE

E. O. Wilson's quote is from p. 139 of Wilson (1984). I'm grateful to the photographers who generously provided their wonderful firefly portraits for this field guide: *Photinus*, *Photuris*, and *Ellychnia* photos are by Croar.net, *Pyractomena angulata* photo is by Stephen Cresswell, *Pyractomena borealis* photo is by Richard Migneault, and *Lucidota atra* photo is by Patrick Coin.

For distinguishing fireflies from other similar-looking beetles, there are several excellent beetle identification guides, including:

White, R. E. (1998). *A Field Guide to the Beetles of North America*. Houghton Mifflin Harcourt, New York, NY.

BugGuide (http://bugguide.net) is a free online resource hosted by Iowa State University. A dedicated team of entomologists provides identifications for insect photos uploaded by curious naturalists from all over the United States and Canada.

Evans, A. V. (2014). *Beetles of Eastern North America*. Princeton University Press, Princeton, NJ.

### Firefly Guides around the World

Field guides to the local firefly fauna have been published for many regions, including Taiwan, Hong Kong, Portugal, China, and Japan. John Day's informative website *Fireflies and Glow-worms* (http://www.firefliesandglow-worms.co.uk/key-to-firefly-genera.html) includes a key to European genera.

Chen T. R. (2003). *The Fireflies of Taiwan*. Field Image Press, Taipei City, Taiwan. 255 pp. (in Chinese).

De Cock, R., H. N. Alves, N. G. Oliveira, and J. Gomes (2015). *Fireflies and Glow-Worms of Portugal (Pirilampos de Portugal)*. Parque Biológico de Gaia, Avintes, Portugal. 80 pp. (in Portuguese and English).

Fu, X. (2014). *Ecological Atlas of Chinese Fireflies*. Commercial Press, Beijing. 167 pp. (in Chinese).

Ohba, N. (2004). *Mysteries of Fireflies* (Hotaru Tenmetsu no Fushigi). Yokosuka City Museum, Yokosuka, Japan (in Japanese).

Vor, Y. (2012). *Fireflies of Hong Kong*. Hong Kong Entomological Society, Hong Kong. 117 pp. (in Chinese).

Surprisingly, until now there haven't been any comprehensive field guides to North American fireflies, though the following resources are quite useful:

*Firefly Watch* (https://legacy.mos.org/fireflywatch). Hosted by the Boston Museum of Science, this site provides descriptions and flash charts for *Photinus*, *Pyractomena*, and *Photuris* lightningbugs.

Faust, L. (2017). *Fireflies, Glow-Worms, and Lightning Bugs! Natural History and a Guide to the Fireflies of the Eastern US and Canada.* University of Georgia Press, Athens.

Lloyd, J. E. (1966). Studies on the flash communication systems of *Photinus* fireflies. *University of Michigan Miscellaneous Publications* No. 130. http://deepblue.lib.umich.edu/handle/2027.42/56374.

Luk, S.P.L., S. A. Marshall, and M. A. Branham (2011). The fireflies (Coleoptera: Lampyridae) of Ontario. *Canadian Journal of Arthropod Identification* 16. http://www.biology.ualberta.ca/bsc/ejournal/lmb_16/lmb_16.html.

Majka, C. G. (2012). The Lampyridae (Coleoptera) of Atlantic Canada. *Journal of the Acadian Entomological Society* 8: 11–29. http://www.acadianes.org/journal.php.

# Glossary

* * *

**aggressive mimicry**: an evolutionary adaptation in which a predator mimics the form or behavior of something harmless in order to gain access to its prey.

**aposematic display**: any combination of sight, sound, or smell produced by noxious prey that warns off a predator *before* it attacks.

**ATP**: adenosine triphosphate, a molecule used to store and then transport energy inside living cells.

**Batesian mimicry**: an evolutionary phenomenon driven by natural selection in which a palatable species gains protection against its predators by copying the warning signal that a noxious species uses for its own protection.

**bioassay**: a test that uses a live animal's response as a readout; such tests are employed, for example, to measure whether a predator is deterred by a specific chemical.

**bufadienolides**: a class of toxic chemicals that many plants and a few animals (toads, fireflies) manufacture to defend themselves against enemies. Originally isolated in 1933 from Egyptian squill, at low doses these steroids can act as cardiac stimulants and antitumor agents.

**Coleoptera** (*Co-lee-**op**-te-rah*): the beetles, which are not only the largest insect order, but also constitute about 25% of all known animal species on Earth. During their lifetimes, all beetles undergo complete metamorphosis, which entails drastic changes to their appearance, habits, and habitats.

***Ellychnia*** (*Ee-**lick**-nee-ah*): a common group of North American dark fireflies closely related to *Photinus* lightningbugs; these adults fly during daytime and don't light up.

**elytra** (*pl.*): the front wings of beetles, which are modified into tough coverings that protect their flight wings.

**enzyme**: a large protein molecule that catalyzes, or speeds up, a particular chemical reaction.

**exaptation**: a characteristic that originally gave its owner a certain advantage that's different from its present-day function.

**Lampyridae** (*Lam-**pier**-ri-dee*): the beetle family to which all fireflies belong.

**larva** (*sing.*), **larvae** (*pl.*): the distinct juvenile stage of insects; fireflies in this life stage are carnivorous and live either underground or underwater.

**lek**: an aggregation where males perform courtship displays, and which females visit to choose their mates (no resources are involved).

**lucibufagins**: toxic chemicals manufactured by some fireflies to defend themselves against predators.

**luciferase**: the generic term used to refer to a group of enzymes that catalyze light production.

**luciferin**: the generic term for a group of light-emitting compounds found in bioluminescent organisms; in a chemical reaction catalyzed by an enzyme (luciferase), luciferin gets transformed into an excited state that emits light.

**mitochondria**: energy factories found within cells of all eukaryotes (comprised of animals, plants, and fungi); these cellular organelles are responsible for manufacturing ATP.

**Müllerian mimicry**: an evolutionary phenomenon driven by natural selection in which two or more noxious species come to look alike by converging on a shared signal that warns off predators.

**natural selection**: an evolutionary process that occurs whenever variation in some inherited characteristic (which could be anatomical, biochemical, physiological, or behavioral) generates differences among individuals in how well they can survive and reproduce.

**nitric oxide (NO)**: a small molecule used to send signals between cells.

**peroxisome**: small organelles located inside cells; in fireflies these are located within the photocytes and house the ingredients for light production.

**pheromone**: a chemical signal that carries information between members of the same species.

***Photinus*** (*Fo-**tie**-nus*): a common group of North American lightningbug fireflies whose males are often eaten by the predatory femmes fatales.

**photocyte**: specialized cells that produce light; in fireflies, these are localized in an organ called the lantern.

***Photuris*** (*Fo-**tur**-es*): a common group of North American lightningbug fireflies that includes the predatory femmes fatales.

**phylogenetics**: the study of evolutionary processes based on inferring historical relationships among groups of living and extinct organisms.

**physiology**: the scientific investigation of how living organisms function, typically studied at the level of cells, organs, or whole organisms.

**pronotum**: the plate-like shield that covers the back of a firefly's head.

**pupa** (*sing.*), **pupae** (*pl.*): the insect life stage that marks the transition between the larval stage and the adult.

**sexual dimorphism**: obvious differences between the males and females of a species in their size or external appearance.

**sexual selection**: a type of natural selection where individuals vary in their ability to attract or compete for mates, or to gain fertilizations.

**unpalatable**: toxic, repellent, or otherwise bad tasting.

# References

* * *


Albo, M. J., G. Winther, C. Tuni, S. Toft, and T. Bilde (2011). Worthless donations: Male deception and female counterplay in a nuptial gift-giving spider. *BMC Evolutionary Biology* 11: 329. http://www.biomedcentral.com/1471-2148/11/329.

Allen, L., and J. Wilson, editors (1992). *Lafcadio Hearn: Japan's Great Interpreter; A New Anthology of His Writings 1894–1904*. Japan Library, Sandgate, Kent, UK.

Andreu, N., et al. (2013). Rapid in-vivo assessment of drug efficacy against *Mycobacterium tuberculosis* using an improved firefly luciferase. *Journal of Antimicrobial Chemotherapy* 68: 2118–27.

Banuls, L.M.Y., E. Urban, M. Gelbcke, F. Dufrasne, B. Kopp, R. Kiss, and M. Zehl (2013). Structure-activity relationship analysis of bufadienolide-induced in vitro growth inhibitory effects on mouse and human cancer cells. *Journal of Natural Products* 76: 1078–84.

Barber, H. S. (1951). North American fireflies of the genus *Photuris*. *Smithsonian Miscellaneous Collection* 117, no. 1. 58 pp.

Bates, H. W. (1862). Contributions to an insect fauna of the Amazon Valley. Lepidoptera: Heliconidae. *Transactions of the Linnaean Society* (London) 23 (3): 495–566. doi:10.1111/j.1096-3642.1860.tb00146.x.

Bauer, C. M., G. Nachman, S. M. Lewis, L. Faust, and J. M. Reed (2013). Modeling effects of harvest on firefly population persistence. *Ecological Modelling* 256: 43–52.

Blum M. S., and A. Sannasi (1974). Reflex bleeding in the lampyrid *Photinus pyralis*: defensive function. *Journal of Insect Physiology* 20: 451–60.

Branham, M. A. (2005). Firefly Communication. pp. 110–12 In: M. Licker, E. Geller, J. Weil, D. Blumel, A. Rappaport, C. Wagner, and R. Taylor (eds.) *The McGraw-Hill 2005 Yearbook of Science and Technology*. McGraw-Hill, New York, NY.

Branham, M. A., and M. Greenfield (1996). Flashing males win mate success. *Nature* 381: 745–46.

Branham M. A., and J. W. Wenzel (2001). The evolution of bioluminescence in cantharoids (Coleoptera: Elateroidea). *Florida Entomologist* 84: 565–86. http://journals.fcla.edu/flaent/article/view/75005.

Branham. M. A., and J. Wenzel (2003). The origin of photic behavior and the evolution of sexual communication in fireflies. *Cladistics* 19: 1–22. http://branhamlab.com/default.asp?action=show_pubs.

Buck, J. B. (1937). *Studies on the Firefly*. PhD thesis, Johns Hopkins University, Baltimore, MD.

Buck, J. B. (1938). Synchronous rhythmic flashing of fireflies. *Quarterly Review of Biology* 13: 301–14. http://www.jstor.org/stable/2808377.

Buck, J. B. (1948). The anatomy and physiology of the light organ in fireflies. *Annals of the New York Academy of Sciences* 49: 397–482.

Buck, J. B. (1988). Synchronous rhythmic flashing of fireflies II. *Quarterly Review of Biology* 65: 265–89. http://www.jstor.org/stable/2830425.

Buck, J. B., and E. Buck (1968). Mechanism of rhythmic synchronous flashing of fireflies. *Science* 159: 1319–27.

Buck, J. B., and E. Buck (1978). Toward a functional interpretation of synchronous flashing by fireflies. *American Naturalist*. 112: 471–92.

Carson, R. (1965). *The Sense of Wonder*. Harper and Row, New York, NY.

Case, J. F. (1980). Courting behavior in a synchronously flashing, aggregative firefly, *Pteroptyx tener*. *Biological Bulletin* 159: 613–25. http://www.biolbull.org/content/159/3/613.

Case, J., and F. Hanson (2004). The luminous world of John and Elisabeth Buck. *Integrative and Comparative Biology* 44: 197–202. http://icb.oxfordjournals.org/content/44/3/197.full.

Cratsley, C. K., and S. M. Lewis (2003). Female preference for male courtship flashes in *Photinus ignitus* fireflies. *Behavioral Ecology* 14: 135–40. http://beheco.oxfordjournals.org/content/14/1/135.full.

Cratsley, C. K., and S. M. Lewis (2005). Seasonal variation in mate choice of *Photinus ignitus* fireflies. *Ethology* 111: 89–100.

Cratsley, C. K., J. Rooney, and S. M. Lewis (2003). Limits to nuptial gift production by male fireflies, *Photinus ignitus*. *Journal of Insect Behavior* 16: 361–70.

Darwin, C. R. (1859). *The Origin of Species by Means of Natural Selection, or the Preservation of Favoured Races in the Struggle for Life*. John Murray, London.

Darwin, C. (1871). *The Descent of Man and Selection in Relation to Sex*. John Murray, London.

Darwin, F., editor (1887). *The Life and Letters of Charles Darwin, Including an Autobiographical Chapter*. John Murray, London.

Davies, N. (1983). Polyandry, cloaca-pecking and sperm competition in dunnocks. *Nature* 302: 334–36.

Day, J. C. (2011). Parasites, predators, and defence of fireflies and glow-worms. *Lampyrid* 1: 70–102.

De Cock, R. (2009). Biology and behaviour of European lampyrids. pp. 161–200. In: V. B. Meyer-Rochow (ed.) *Bioluminescence in Focus—A Collection of Illuminating Essays*. Research Signpost, Kerala, India.

De Cock, R., and E. Matthysen (2001). Do glow-worm larvae (Coleoptera: Lampyridae) use warning coloration? *Ethology* 107: 1019–33.

De Cock R., and E. Matthysen (2003). Glow-worm larvae bioluminescence (Coleoptera: Lampyridae) operates as an aposematic signal upon toads (*Bufo bufo*). *Behavioral Ecology* 14: 103–8. http://beheco.oxfordjournals.org/content/14/1/103.full.

De Cock R., and E. Matthysen (2005). Sexual communication by pheromones in a firefly, *Phosphaenus hemipterus* (Coleoptera: Lampyridae). *Animal Behaviour* 70: 807–18.

De Cock, R., L. Faust, and S. M. Lewis (2014). Courtship and mating in *Phausis reticulata* (Coleoptera: Lampyridae): Male flight behaviors, female glow displays, and male attraction to light traps. *Florida Entomologist* 97: 1290–1307. http://dx.doi.org/10.1653/024.097.0404.

Demary, K., C. Michaelidis, and S. M. Lewis (2006). Firefly courtship: Behavioral and morphological predictors of male mating success in *Photinus greeni*. *Ethology* 112: 485–92.

Eberhard, W. G. (1996). *Female Control: Sexual Selection by Cryptic Female Choice*. Princeton University Press, Princeton, NJ. 472 pp.

Eisner, T. (2003). *For Love of Insects*. Belknap Press of Harvard University Press, Cambridge, MA. 464 pp.

Eisner, T., M. A. Goetz, D. E. Hill, S. R. Smedley, and J. Meinwald (1997). Firefly "femme fatales" acquire defensive steroids (lucibufagins) from their firefly prey. *Proceedings of the National Academy of Sciences USA* 94: 9723–28.

Eisner, T., and J. Meinwald (1995). The chemistry of sexual selection. *Proceedings of the National Academy of Sciences USA* 92: 50–55. http://www.ncbi.nlm.nih.gov/pmc/articles/PMC42815.

Eisner, T., D. F. Wiemer, L. W. Haynes, and J. Meinwald (1978). Lucibufagins: Defensive steroids from the fireflies *Photinus ignitus* and *P. marginellus* (Coleoptera: Lampyridae). *Proceedings of the National Academy of Sciences USA* 75: 905–8. http://www.pnas.org/content/75/2/905.full.pdf.

Faust, L. (2010). Natural history and flash repertoire of the synchronous firefly *Photinus carolinus* in the Great Smoky Mountains National Park. *Florida Entomologist* 93: 208–17. http://journals.fcla.edu/flaent/article/view/76082.

Faust, L. (2012). Fireflies in the snow: Observations on two early-season arboreal fireflies *Ellychnia corrusca* and *Pyractomena borealis*. *Lampyrid* 2: 48–71.

Faust, L., S. M. Lewis, and R. De Cock (2012). Thieves in the night: Kleptoparasitism by fireflies in the genus *Photuris* (Coleoptera: Lampyridae). *Coleopterists Bulletin* 66: 1–6.

Fender, K. M. (1970). *Ellychnia* of western North America (Coleoptera: Lampyridae). *Northwest Science* 44: 31–43.

Fisher, R. A. (1930). *The Genetical Theory of Natural Selection*. Clarendon Press, Oxford.

Frick-Ruppert, J., and J. Rosen (2008). Morphology and behavior of *Phausis reticulata* (Blue Ghost Firefly). *Journal of North Carolina Academy of Science* 124: 139–47. http://dc.lib.unc.edu/cdm/ref/collection/jncas/id/3883.

Fu, X., V. Meyer-Rochow, J. Tyler, H. Suzuki, and R. De Cock (2009). Structure and function of the eversible organs of several genera of larval firefly (Coleoptera: Lampyridae). *Chemoecology* 19: 155–68.

Fu, X., F. Vencl, N. Ohba, V. Meyer-Rochow, C. Lei, and Z. Zhang (2007). Structure and function of the eversible glands of the aquatic firefly *Luciola leii* (Coleoptera: Lampyridae). *Chemoecology* 17: 117–24.

Gao, H., R. Popescu, B. Kopp, and Z. Wang (2011). Bufadienolides and their antitumor activity. *Natural Product Reports* 28: 953–69. http://dx.doi.org/10.1039/c0np00032a.

Ghiradella, H. (1998). Anatomy of light production: The firefly lantern. In: F. W. Harrison and M. Locke (eds.) *Microscopic Anatomy of Invertebrates*, Volume 11A: *Insecta*. Wiley-Liss, New York, NY.

Ghiradella, H., and J. T. Schmidt (2008). Fireflies: Control of flashing. pp. 1452–63. In: J. Capinera (ed.) *Encyclopedia of Entomology*. Springer, New York, NY.

González, A., J. F. Hare, and T. Eisner (1999). Chemical egg defense in *Photuris* firefly "femmes fatales." *Chemoecology* 9: 177–85.

Goodenough, U. (1998). *The Sacred Depths of Nature*. Oxford University Press, New York, NY. 224 pp.

Green, J. W. (1956). Revision of the Nearctic species of *Photinus* (Coleoptera: Lampyridae). *Proceedings of the California Academy of Sciences* Series 4, Vol. 28: 561–13.

Green, J. W. (1957). Revision of the Nearctic species of *Pyractomena* (Coleoptera: Lampyridae). *Wasmann Journal of Biology* 15: 237–84.

Greenfield, M. (2002). *Signalers and Receivers: Mechanisms and Evolution of Arthropod Communication*. Oxford University Press, New York, NY.

Gronquist, M., F. C. Schroeder, H. Ghiradella, D. Hill, E. M. McCoy, J. Meinwald, and T. Eisner (2006). Shunning the night to elude the hunter: Diurnal fireflies and the "femmes fatales." *Chemoecology* 16: 39–43.

Gwynne D. T., and D. Rentz (1983). Beetles on the bottle: Male buprestids mistake stubbies for females (Coleoptera). *Australian Journal of Entomology* 22: 79–80. http://onlinelibrary.wiley.com/doi/10.1111/j.1440-6055.1983.tb01846.x/pdf.

Healya, K., L. McNally, G. D. Ruxton, N. Cooper, and A. Jackson (2013). Metabolic rate and body size are linked with perception of temporal information. *Animal Behaviour* 86: 685–96. http://dx.doi.org/10.1016/j.anbehav.2013.06.018.

Hosoe, T., K. Saito, M. Ichikawa, and N. Ohba (2014). Chemical defense in the firefly *Rhagophthalmus ohbai* (Coleoptera: Rhagophthalmidae). *Applied Entomology and Zoology* 49: 331–35.

Iguchi, Y. (2009). The ecological impact of an introduced population on a native population in the firefly *Luciola cruciata* (Coleoptera: Lampyridae). *Biodiversity and Conservation* 18: 2119–26. http://link.springer.com/article/10.1007%2Fs10531-009-9576-8.

Ineichen, S., and B. Rüttimann (2012). Impact of artificial light on the distribution of the common European glow-worm, *Lampyris noctiluca*. *Lampyrid* 2: 31–36.

Jusoh, W. F. A. W., and N. R. Hashim (2012). The effect of habitat modification on firefly populations at the Rembau-Linggi estuary, Peninsular Malaysia. *Lampyrid* 2: 149–55.

Kawahara, A. (2007). Thirty-foot telescopic nets, bug-collecting video games, and beetle pets: Entomology in modern Japan. *American Entomologist* 53: 160–72. http://dx.doi.org/10.1093/ae/53.3.160.

Knight, M., R. Glor, S. R. Smedley, A. González, K. Adler, and T. Eisner (1999). Firefly toxicosis in lizards. *Journal of Chemical Ecology* 25 (9): 1981–86.

Koene, J. (2006). Tales of two snails: Sexual selection and sexual conflict in *Lymnaea stagnalis* and *Helix aspersa*. *Integrative and Comparative Biology* 46: 419–29. http://intl-icb.oxfordjournals.org/content/46/4/419.full.

Laurent, E. (2001). Mushi. *Natural History* (March): 70–75.

Lee, K., Y. Kim, J. Lee, M. Song, and S. Nam (2008). Toxicity to firefly, *Luciola lateralis*, of commercially registered insecticides and fertilizers. *Korean Journal of Applied Entomology* 47: 265–72 (in Korean).

Lewis, S. M. (2009). Bioluminescence and sexual signaling in fireflies. In: V. B. Meyer-Rochow (ed.) *Bioluminescence in Focus—A Collection of Illuminating Essays*. Research Signpost, Kerala, India.

Lewis, S. M., and C. K. Cratsley (2008). Flash signal evolution, mate choice and predation in fireflies. *Annual Review of Entomology* 53: 293–321.

Lewis, S. M., C. K. Cratsley, and J. A. Rooney (2004). Nuptial gifts and sexual selection in *Photinus* fireflies. *Integrative and Comparative Biology* 44: 234–37. http://intl-icb.oxfordjournals.org/content/44/3/234.full.

Lewis, S. M., L. Faust, and R. De Cock (2012). The dark side of the Light Show: Predation on fireflies of the Great Smokies. *Psyche*. http://dx.doi.org/10.1155/2012/634027.

Lewis, S. M., and A. South (2012). The evolution of animal nuptial gifts. *Advances in the Study of Behavior* 44: 53–97.

Lewis, S. M., K. Vahed, J. M. Koene, L. Engqvist, L. F. Bussière, J. C. Perry, D. Gwynne, and G.U.C. Lehmann (2014). Emerging issues in the evolution of animal nuptial gifts. *Biology Letters* 10: 20140336. http://dx.doi.org/10.1098/rsbl.2014.0336.

Lewis, S. M., and O. Wang (1991). Reproductive ecology of two species of *Photinus* fireflies (Coleoptera: Lampyridae). *Psyche* 98: 293–307. http://www.hindawi.com/journals/psyche/1991/076452/abs/.

Lloyd, J. E. (1965). Aggressive mimicry in *Photuris*: Firefly *femmes fatales*. *Science* 149: 653–54.

Lloyd, J. E. (1966). Studies on the flash communication system in *Photinus* fireflies. *University of Michigan Miscellaneous Publications* 130: 1–95. http://deepblue.lib.umich.edu/handle/2027.42/56374.

Lloyd, J. E. (1972). Chemical communication in fireflies. *Environmental Entomology* 1: 265–66.

Lloyd, J. E. (1973a). Firefly parasites and predators. *Coleopterists Bulletin* 27: 91–106. http://www.jstor.org/discover/10.2307/3999442.

Lloyd, J. E. (1973b). Model for the mating protocol of synchronously flashing fireflies. *Nature* 245: 268–70.

Lloyd, J. E. (1975). Aggressive mimicry in *Photuris* fireflies: Signal repertoires by *femmes fatales*. *Science* 187: 452–53.

Lloyd J. E. (1979a). Sexual selection in luminescent beetles. pp. 293–342. In: M. S. Blum and N. A. Blum (eds.) *Sexual Selection and Reproductive Competition in Insects*. Academic Press, New York, NY. 463 pp.

Lloyd, J. E. (1979b). Symposium: Mating behavior and natural selection. *Florida Entomologist* 62: 17–34.

Lloyd, J. E. (1984). On the occurrence of aggressive mimicry in fireflies. *Florida Entomologist* 67: 368–76. http://journals.fcla.edu/flaent/article/view/57933.

Lloyd, J. E. (2000). On research and entomological education IV: Quantifying mate search in a perfect insect-seeking true facts and insight (Coleoptera: Lampyridae, *Photinus*) *Florida Entomologist* 83: 211–28. http://journals.fcla.edu/flaent/article/view/59545.

Lloyd, J. E. (2002). Family 62: Lampyridae. pp. 187–96. In: R. H. Arnett, M. C. Thomas, P. E. Skelley, and J. H. Frank (eds.) *American Beetles*, Volume II: *Polyphaga: Scarabaeoidea through Curculionoidea*. CRC Press, Boca Raton, FL.

Lloyd, J. E. (2008). Fireflies (Coleoptera: Lampyridae). pp 1429–52. In: J. L. Capinera (ed.) *Encyclopedia of Entomology*. Springer, New York, NY.

Lloyd, J. E., and S. Wing (1983). Nocturnal aerial predation of fireflies by light-seeking fireflies. *Science* 222: 634–35.

Long, S. M., S. Lewis, L. Jean-Louis, G. Ramos, J. Richmond, and E. M. Jakob (2012). Firefly flashing and jumping spider predation. *Animal Behaviour* 83: 81–86.

Lynch, V. J. (2007). Inventing an arsenal: Adaptive evolution and neofunctionalization of snake venom phospholipase A2 genes. *BMC Evolutionary Biology* 7 (2). http://www.biomedcentral.com/1471-2148/7/2.

Majka, C. G., and J. S. MacIvor (2009). The European lesser glow worm, *Phosphaenus hemipterus*, in North America (Coleoptera, Lampyridae). *ZooKeys* 29: 35–47. doi: 10.3897/zookeys.29.279.

Maurer, U. (1968). Some parameters of photic signalling important to sexual and species recognition in the firefly *Photinus pyralis*. Unpublished master's thesis. State University of New York, Stony Brook. 114 pp.

McDermott, F. (1964). The taxonomy of the Lampyridae. *Transactions of the American Entomological Society* 90: 1–72. http://www.jstor.org/stable/25077867.

McDermott, F. A. (1967). The North American fireflies of the genus *Photuris* Dejean: A modification of Barber's key (Coleoptera; Lampyridae). *Coleopterists Bulletin* 21: 106–16. http://www.jstor.org/stable/3999313.

McKenna, D., and B. Farrell (2009). Beetles (Coleoptera). pp. 278–89. In: S. B. Hedges and S. Kumar (eds.) *The Timetree of Life*. Oxford University Press, New York, NY. 572 pp.

Michaelidis, C., K. Demary, and S. M. Lewis (2006). Male courtship signals and female signal assessment in *Photinus greeni* fireflies. *Behavioral Ecology* 17: 329–35. http://beheco.oxfordjournals.org/content/17/3/329.full.

Moiseff, A., and J. Copeland (1995). Mechanisms of synchrony in the North American firefly *Photinus carolinus*. *Journal of Insect Behavior* 8: 395–407.

Moiseff, A., and J. Copeland (2010). Firefly synchrony: A behavioral strategy to minimize visual clutter. *Science* 329: 181.

Moosman, P., C. K. Cratsley, S. D. Lehto, and H. H. Thomas (2009). Do courtship flashes of fireflies (Coleoptera: Lampyridae) serve as aposematic signals to insectivorous bats? *Animal Behaviour* 78: 1019–25.

Nada, B., L. G. Kirton, Y. Norma-Rashid, and V. Khoo (2009). Conservation efforts for the synchronous fireflies of the Selangor River in Malaysia. pp. 160–71. In: B. Napompeth (ed.) *Proceedings of the 2008 International Symposium on Diversity and Conservation of Fireflies*. Queen Sirikit Botanic Garden, Chiang Mai, Thailand.

Niwa, K., Y. Ichino, and Y. Ohmiya (2010). Quantum yield measurements of firefly bioluminescence using a commercial luminometer. *Chemical Letters* 39: 291–3.

Oba, Y. (2015). Insect bioluminescence in the post-molecular biology era. pp. 94–119. In: K. H. Hoffmann (ed.) *Insect Molecular Biology and Ecology*. CRC Press, Boca Raton, FL.

Oba, Y., M. Branham, and T. Fukatsu (2011). The terrestrial bioluminescent animals of Japan. *Zoological Science* 28: 771–89.

Oba, Y., T. Shintani, T. Nakamura, M. Ojika, and S. Inouye (2008). Determination of the luciferin content in luminous and non-luminous beetles. *Bioscience Biotechnology and Biochemistry* 72: 1384–87.

Ohba, N. (2004). *Mysteries of Fireflies*. Yokosuka City Museum, Yokosuka, Japan (in Japanese).

Peretti, A. V., and A. Aisenberg, editors. (2015). *Cryptic Female Choice in Arthropods: Patterns, Mechanisms, and Prospects*. Springer International, London. 509 pp.

Pieribone, V., and D. Gruber (2005). A Glow in the Dark: The Revolutionary Science of Biofluorescence. Harvard University Press, Cambridge, MA.

Pietsch, T. (2005). Dimorphism, parasitism, and sex revisited: Modes of reproduction among deep-sea ceratioid anglerfishes (Teleostei: Lophiiformes). *Ichthyological Research* 52: 207–36.

Pizzari, T., and N. Wedell (2013). Introduction: The polyandry revolution. *Philosophical Transactions of the Royal Society B* 368: 20120041.

Rich, C., and T. Longcore, editors (2006). *Ecological Consequences of Artificial Night Lighting.* Island Press, Washington, DC. 480 pp.

Rooney, J., and S. M. Lewis (1999). Differential allocation of male-derived nutrients in two lamyprid beetles with contrasting life-history characteristics. *Behavioral Ecology* 10: 97–104. http://beheco.oxfordjournals.org/content/10/1/97.full.

Rooney, J. A., and S. M. Lewis (2000). Notes on the life history and mating behavior of *Ellychnia corrusca* (Coleoptera: Lampyridae). *Florida Entomologist* 83: 324–34. http://journals.fcla.edu/flaent/article/view/59556.

Rooney, J., and S. M. Lewis (2002). Fitness advantage of nuptial gifts in female fireflies. *Ecological Entomology* 27: 373–77.

Rosellini, D. (2012). Selectable markers and reporter genes: A well-furnished toolbox for plant science and genetic engineering. *Critical Reviews in Plant Sciences* 31: 401–53.

Simmons, L. W. (2001). *Sperm Competition and Its Evolutionary Consequences in the Insects.* Princeton University Press, Princeton, NJ. 456 pp.

Smith, H. M. (1935). Synchronous flashing of fireflies. *Science* 82: 151–52. http://www.sciencemag.org/content/82/2120/151.short.

South, A., and S. M. Lewis (2012a). Determinants of reproductive success across sequential episodes of sexual selection in a firefly. *Proceedings of the Royal Society B* 279: 3201–8. http://dx.doi.org/10.1098/rspb.2012.0370.

South, A., and S. M. Lewis (2012b). Effects of male ejaculate on female reproductive output and longevity in *Photinus* fireflies. *Canadian Journal of Zoology* 90: 677–81.

South, A., T. Sota, N. Abe, M. Yuma, and S. M. Lewis (2008). The production and transfer of spermatophores in three Asian species of *Luciola* fireflies. *Journal of Insect Physiology* 54: 861–66.

South, A., K. Stanger-Hall, M. Jeng, and S. M. Lewis (2011). Correlated evolution of female neoteny and flightlessness with male spermatophore production in fireflies (Coleoptera: Lampyridae). *Evolution* 65: 1099–113.

Stanger-Hall, K., D. Hillis, and J. Lloyd (2007). Phylogeny of North American fireflies: Implications for the evolution of light signals. *Molecular Phylogenetics and Evolution* 45: 33–39.

Strogatz, S. H. (2003). *Sync: The Emerging Science of Spontaneous Order*. Hyperion Books, New York, NY. 338 pp.

Thancharoen, A. (2012). Well-managed firefly tourism: A good tool for firefly conservation in Thailand. *Lampyrid* 2: 142–48.

Tinbergen, N. (1963). On aims and methods of ethology. *Zeitschrift für Tierpsychologie* 20: 410–33.

Trimmer, B. A., J. R. Aprille, D. Dudzinski, C. Lagace, S. M. Lewis, T. Michel, S. Qazi, and R. Zayas (2001). Nitric oxide and the control of firefly flashing. *Science* 292: 2486–88.

Trivers, R. (1972). Parental investment and sexual selection. pp. 136–79. In: B. Campbell (ed.) *Sexual Selection and the Descent of Man, 1871–1971*. Aldine, Chicago.

Tyler, J. (2002). *The Glow-worm*. Privately published.

Tyler, J., W. McKinnon, G. Lord, and P. J. Hilton (2008). A defensive steroidal pyrone in the glow-worm *Lampyris noctiluca* L. (Coleoptera: Lampyridae). *Physiological Entomology* 33: 167–70.

Underwood, T. J., D. W. Tallamy, and J. D. Pesek (1997). Bioluminescence in firefly larvae: A test of the aposematic display hypothesis (Coleoptera: Lampyridae). *Journal of Insect Behavior* 10: 365–70.

van der Reijden, E., J. Monchamp, and S. M. Lewis (1997). The formation, transfer, and fate of male spermatophores in *Photinus* fireflies (Coleoptera: Lampyridae). *Canadian Journal of Zoology* 75: 1202–5.

Vencl, F. V., and A. D. Carlson (1998). Proximate mechanisms of sexual selection in the firefly *Photinus pyralis* (Coleoptera: Lampyridae). *Journal of Insect Behavior* 11: 191–207.

Viviani, V. (2001). Fireflies (Coleoptera: Lampyridae) from southeastern Brazil: Habitats, life history, and bioluminescence. *Annals of the Entomological Society of America* 94: 129–45.

Viviani, V. (2002). The origin, diversity, and structure function relationships of insect luciferases. *Cellular and Molecular Life Sciences* 59: 1833–50.

von Uexküll, J. (1934). A stroll through the worlds of animals and men: A picture book of invisible worlds. pp. 5–80. In: C. H. Schiller (ed.) *Instinctive Behavior: The Development of a Modern Concept*. International Universities Press, New York, NY.

Waage, J. K. (1979). Dual function of the damselfly penis: Sperm removal and transfer. *Science* 203: 916–18.

Wallace, A. R. (1867). Mimicry, and other protective resemblances among animals. *Westminster Review* 88: 1–20.

Wallace, A. R. (1889). *Darwinism—An Exposition of the Theory of Natural Selection with Some of Its Applications.* Macmillan, London.

Williams, F. X. (1917). Notes on the life-history of some North American Lampyridae. *Journal of the New York Entomological Society* 25: 11–33. http://www.jstor.org/stable/25003739.

Wilson, E. O. (1984). *Biophilia*. Harvard University Press, Cambridge, MA.

Wilson, T., and J. W. Hastings (1998). Bioluminescence. *Annual Review of Cell and Developmental Biology* 14: 197–230.

Wilson, T., and J. W. Hastings (2013). *Bioluminescence: Living Lights, Lights for Living*. Harvard University Press, Cambridge, MA. 208 pp.

Wing, S., J. E. Lloyd, and T. Hongtrakul (1982). Male competition in *Pteroptyx* fireflies: Wing-cover clamps, female anatomy, and mating plugs. *Florida Entomologist* 66: 86–91. http://journals.fcla.edu/flaent/article/view/57785.

Woods W. A., H. Hendrickson, J. Mason, and S. M. Lewis (2007). Energy and predation costs of firefly courtship signals. *American Naturalist* 170: 702–8.

Yoshizawa, K., R. L. Ferreira, Y. Kamimura, and C. Lienhard (2014). Female penis, male vagina, and their correlated evolution in a cave insect. *Current Biology* 24: 1006–10.

Yuma, M. (1993). *Hotaru no mizu, hito no mizu* (Fireflies' water, human's water). Shinhyoron, Tokyo (in Japanese).

Yuma, M. (1995). The welfare of Moriyama fireflies. *Japan Association for Firefly Research* 28: 29–31 (in Japanese).

Yuma, M., and M. Hori (1981). Gregarious oviposition of *Luciola cruciata*. *Physiology and Ecology Japan* 181: 93–112.

# Index

* * *

Italic pages refer to figures